OLD WINE
NEW FLASKS

OLD WINE

Reflections on Science

and Jewish Tradition

NEW FLASKS

Roald Hoffmann

Shira Leibowitz Schmidt

W. H. FREEMAN AND COMPANY
NEW YORK

Text and Cover Designer: Blake Logan

Library of Congress Cataloging–in–Publication Data

Hoffmann, Roald
    Old wine, new flasks: reflections on science and Jewish tradition
    Roald Hoffmann, Shira Leibowitz Schmidt.
        p.   cm.
    Includes index.
    ISBN 0-7167-2899-0
    1. Judaism and science.   I. Schmidt, Shira Leibowitz.   II. Title.
BM538.S3H64   1997
296.3'75--DC21   97-2706
                                                                      CIP
                                                                      r97

Printed in the United States of America

First printing, 1997

Rabbi Meir says:

Look not at the flask but at what it contains.
There may be a new flask full of old wine,
and an old flask that has not even new wine in it.

Mishnah Avot 4:29

רַבִּי מֵאִיר אוֹמֵר: אַל תִּסְתַּכֵּל בַּקַּנְקַן אֶלָּא בְּמַה שֶׁיֵּשׁ בּוֹ;

יֵשׁ קַנְקַן חָדָשׁ מָלֵא יָשָׁן,

וְיָשָׁן שֶׁאֲפִלּוּ חָדָשׁ אֵין בּוֹ.

Contents

Preface   ix

1 Is Nature Natural?   1

2 A Sukkah *from an Elephant*   55

3 You Must Not Deviate to the Right or the Left   79

4 Bitter Waters Run Sweet   123

5 The Flag That Came out of the Blue: A Play in Three Acts and Two Intermezzi   159

6 Signs and Portents: No Parking in the Courtroom   213

7 Pure/Impure   239

8 Camel Caravans in the Pentagon   263

Epilogue   291

How We Came to *Old Wine, New Flasks,* with a Little Help From Our Friends   293

Notes   303

Credits   349

Glossary of Hebrew and Yiddish Terms   351

Index   357

Preface

1

IT WAS A TIE; THE HEAVENLY VOTE was split right down the middle—two in favor, two against. At issue—"Should man be created?" The ministering angels formed parties: Love said, "Yes, let him be created, because he will dispense acts of love"; while Truth argued, "No, let him not be created, for he is a complete fake." Righteousness countered, "Yes, let him be created, because he will do righteous deeds"; and Peace demurred, "Let him not be created, for he is one mass of contention." The score was even: Love and Righteousness in favor; Truth and Peace against.

What did the Lord do? He took Truth and hurled it to the ground, smashing it into thousands of jagged pieces. Thus He broke the tie. Now, two to one in favor, man was created. The ministering angels dared to ask the Master of the Universe, "Why do You break Your emblem, Truth?" for indeed Truth was His seal and emblem. He answered, "Let truth spring up from the earth." (Psalm 85:11). From then on truth was dispersed, splintered into fragments, like a jigsaw puzzle. While a person might find a piece, it held little meaning until he joined with others who had painstakingly gained different pieces of the puzzle. Only then, slowly and deliberately, could they try to fit their pieces of Truth together. To make sense, some sense of things.[1]

2

This ancient Jewish allegory strikes a chord with the authors of *Old Wine, New Flasks,* as they attempt to put together the (all too few) jigsaw pieces they have found. That science and religion only contend, or that they occupy separate compartments in our minds, one unrelated to the other—these are both such impoverishing views. Scientific knowledge, faith, and aesthetics cohabit. They speak to one another in the human soul—yes, sometimes their dialogue is uneasy. But it is their intertwined voices that shape true human understanding.

This metaphor is explored in *Old Wine, New Flasks*. The religious setting is that of Jewish tradition. The science is mainly the central one, chemistry, but biology, physics, and geology speak as well. The artistic context is primarily that of painting, poetry, and music. We tell stories, inherently digressive the way real life is, of how science and religion (and occasionally art) look at pieces of the world.

3

Why specifically chemistry, and why Judaism? No question, we are inclined to what we know best. But we also think these are particularly well suited for exploring the interaction of science and religion.

Chemistry is the craft, art, business, and finally science, of substances and their transformations. With great ingenuity and much labor, we have learned that the substances are, deep within, molecular. Though small, these persistent groupings of atoms are nicely positioned in the middle—not the ultimately small quark, not the immense galactic cluster. Molecules are intermediate—just where human beings are. . . . Which is why human beings have ambiguous attitudes toward chemicals—these molecular groupings of which we and everything else are made can really heal, and really hurt. Chemistry is on the human scale.

So is Jewish religious tradition. Not that other religious structures (diverse as only the human condition can make them) are of less spiritual value to their believers. Except for our relative ignorance of them, we could have used more examples from other religions. But there is something special in the Bible, Talmud, and rabbinical responsa. This is the emphasis on everyday life and the way that life interacts with the spiritual realm. Look at a page of the Talmud, and what do you see? Not the soaring heights of philosophical discourse, but a meticulous, detailed discussion of the compensation for loss and injury if someone stumbles over and breaks a wine flask you left in an unloading zone.

Jewish religious law is called *halakhah*. The word is old, and shares its root with that of the verb "to walk,"or "to proceed." Indeed one way to describe *halakhah* is that to the observant Jew the law indeed is "the way" to follow, again very much in the simple everyday way that we wend through life.

Thus the concern of Jewish religious texts is often *mundane* in the true sense of the word. What about the concern of science? Despite occasional claims to the contrary, we think science is common sense, mathematicized. Good science is looking at the real things of the everyday world, the *mundus* that surrounds us. We do so with ingenious tools, with measurements—with cool (and hot) thought trying to understand the world.

No surprise then that the concerns of Jewish religious inquiry often intersect those of science. In this book we will encounter the rabbi of Atlanta, the dean of the Orthodox Rabbinate in the South, as he pores over the ingredients of Coca-Cola, lest a ritually forbidden component be therein. A concern for the composition of an unknown mixture (yes, even Coca-Cola) is the stuff of analytical chemistry. We will encounter a *hasidic* rabbi in nineteenth-century Russia journeying to Italy to find the shellfish source of a blue dye, the secret of whose production was lost to the Jews some thirteen hundred years before. And tell how that dye was central to a heretical rebellion, a biblical tale that resonates today.

Simple, material things, and the problems they pose for scientific and religious consideration, will lead us in this book to pivotal philosophical and societal issues—of the nature of Nature; of purity and its lack; of perception and representation; of the character of rebellion; of ecological concerns; of signs (and icons) of the times; of the sacred and the profane; of right-handedness as a metaphor for authority (scientific and religious) claimed, contested, and conferred. It's amazing where thinking about how your left hand differs from your right can lead you!

4

We do not wish to minimize the differences between science and religion. In the spectrum of being the same and not the same, science and religion are perhaps more different than alike. And, at times, in every culture, they've been led into conflict—be it the American textbook wars on creationism; Galileo and the Catholic Church in the seventeenth century; some ultraorthodox versus archaeologists in Israel. But we believe that science and Jewish religious tradition share this: *the conviction that this world is very much real and tangible, that the world and the actions of human beings matter, and that there is order to be found.* This commonality is a lot to build on.

There is no need to look for fuzzy systems of religious thought supposedly more amenable to scientific thinking, nor to wring one's hands at the "irrational" persistence of serious religious conviction after two centuries of science. Nor will one find in science a "justification" for what is in the Bible. We believe that the middle ground is just there, to be found in the jigsaw puzzle pieces, the richness and complexity of human beings and the world. That middle ground is reached if one respects all the ways that human beings have devised for trying to understand this world.[2]

5

One final comment before we set out: The religious convictions and expertise of the authors of this book differ. As a consequence, *Old Wine, New Flasks* explores a dialogic, contrapuntal form. Many of the chapters are exchanges or debates. In this way the underlying tension of the agnostic and the religious is exposed, and two faces of contemporary man and woman come into view. The authors contend, in structured harmony (and not taking themselves too seriously), with a shared respect for the underlying unity of all knowledge.

Such contention is actually a time-honored way of Jewish debate (of which more, anon). And, in another way, such debate is at the heart of science.

Is Nature Natural?

4 Tishrei (Sept. 28)
Kiryat Sanz, Netanya, Israel

Dear Ayyal,

Where are you? This past summer you told us that you had decided to spend some time studying in a *yeshivah** in Jerusalem, and postpone your freshman year at Harvard. We'd like you to come here for the Sukkot holiday, if you are in Israel.

Where are you?

Shira and Baruch Schmidt

Oct. 19
Ship Rock, New Mexico, USA

Dear Shira and Baruch,

My parents forwarded your letter to me. You asked twice, "Where are you?" so I will answer you twice. (1) Geographically, I am neither at Harvard nor in the Jerusalem *yeshivah*. I am working in New Mexico, learning real organic farming from Native Americans. (2) Where am I spiritually? For the time being, I don't

---

* A Jewish religious academy, usually for higher studies. There is a glossary of Hebrew and Yiddish terms at the end of the book.

want to go on in science; neither do I want to study at a *yeshivah*. To state my view simplistically, tendentiously, and provocatively— I now see that science/technology and religion have together conspired to destroy this green planet.

The God of Genesis is to blame, for didn't He say to man and woman:

> Be fruitful and multiply, and replenish the earth, and subdue it: and have dominion over the fish of the sea, and over the fowl of the air, and over every living thing that moveth upon the earth.[1]

And science is to blame, that gang of mostly white, Western men, ostensibly seeking wisdom, cuts up Nature and transforms it, screaming of their vaunted neutrality all the way to a polluted hell. Science and religion—never mind the supposed enmity of the two— work together to ruin Nature. For science and religion both assume the world is there for human beings to probe, dominate, rule, and exploit.

I've called a time out from the fast track to take stock of myself and my place in Nature. I have a lot to learn from the Navajo here, who say that when a person lives in harmony with the land and its spirit he is walking in beauty. The landscape is numinous—take the Ship Rock, sacred to the Navajo, an eroded, fractured volcanic rock thrusting 1,800 feet heavenward from the desert, like a schooner in full sail (the real name is Tse Bitai, winged rock). The land and its people are teaching me something of possibilities . . . and limits.

Please, Baruch and Shira, give my regards to your kids—to Shamai, Nathaniel, Tirza, Akiva, Efrat, Michael, and Rahel. I guess you've taken that verse from Genesis (be fruitful and multiply) pretty seriously. . . .

Ayyal

10 MarHeshvan (Nov. 3)
Netanya

Dear Ayyal,

Wow! It's hard to know where to begin to answer your double calumny. Maybe with a glass of cognac (distilled from organic grapes, and kosher, of course).

As with any provocation, yours has some elements of truth in it. But just as the desert—New Mexico's or Israel's—is not just a sea of sand, but a living, complicated ecosystem, so the question of what science and religion have to do with nature (or what they have done to nature) is not as simple as you make it out to be. I have thought a lot about this, believe me. I've spoken with our children, friends, and with other idealistic young men like you. So, could we talk about it, Ayyal? Through letters—we all love your letters.

I begin with some humility—I don't have the answers. You also know, however, that I like a rational analysis of any problem. The position you take is a strong one. There is emotion and reason behind it, and if I am to agree or disagree with it, I must think my way through—with your help—to the premises, the questions that underlie your argument.

I see at least five distinct questions:

1. What is Nature—to science, to religion, to human beings?
2. What is the difference—materially, chemically—between the natural and the synthetic?
3. What is the difference—spiritually, poetically, even erotically—between natural and synthetic?
4. Do scientific knowledge and technological progress inevitably harm nature?
5. Is the biblical imperative really for humanity to dominate nature?

Let's deal with these one by one.

You also know that I like to tell people what to do. So, to begin, I would ask you to find a biblical passage about nature, one that you like.

I'll tell you why later. . . .

Shira

Nov. 17
Ship Rock, N.M.

Dear Shira,

It sounds as if you are handing me an outline for a paper I should write. But, OK, let me try.

Here's the passage I would choose, God speaking to Job out of the whirlwind:

> *By what path is the west wind dispersed,*
> *The east wind scattered over the earth?*
> *Who cut a channel for the torrents*
> *And a path for thunderstorms,*
> *To rain down on uninhabited land,*
> *On the wilderness where no man is,*
> *To saturate the desolate wasteland,*
> *And make the crop of grass sprout forth?*
> *Does the rain have a father?*
> *Who begot the dewdrops?*
> *From whose belly came forth the ice?*
> *Who gave birth to the frost of heaven?*
> *Water congeals like a stone,*
> *And the surface of the deep compacts.*
> *Can you tie cords to Pleiades*
> *Or undo the reins of Orion?*[2]

Ayyal

7 Kislev (Nov. 30)
Netanya

Dear Ayyal,

You've chosen well. In this, one of the greatest of all poems, God speaks to Job of His power and creation. And God indeed uses as exemplars of His power a series of awesome, spellbinding images from *nature*. To God's speech the only answer possible is Job's: "See, I am of small worth; what can I answer You?"

This passage from the Bible, this rephrasing of Creation, is *so* firmly rooted in the imagery of nature. Nature is essential to the dialogue of God with human beings, it is the way God makes man understand. But, Ayyal—Where is the word for nature in this?

The modern Hebrew word for nature, *teva*, is ubiquitous. As I write you, I am shod with *sandaley teva*, Israel's answer to Birkenstock; I am preparing a vegetarian *tivoni* dinner of kosher soya cheeseburger (*tivol*), to be washed down with natural spring water (*mei teva*). The waters may have come from the bucolic Galilean village of Tivon. I've even collected (it didn't take much work) a number of ads in Hebrew and English that invoke nature. [See Plate 1 and illustration on page 6.]

So here's all this nature around us. Being used, I want you to note carefully, not in the service of ecology, but by advertisers out to make money.

Now . . . can you find that word *teva* in the Bible, or in the Talmud? That's all for now—company's coming.

Shira

Dec. 10
Ship Rock, N.M.

Dear Shira,

I have been searching high and low for Nature in the Bible. I tried the Creation story, and found plenty of heavenly bodies, fauna, and flora, but not the Hebrew word for Nature. I tried the Psalms also replete with rapturous descriptions of nature, but no *teva*.

I made my way to the Navajo college library and tried an English Bible concordance—no Nature there either. I got so frustrated I called home to my rabbi and set him searching. He said the closest he could come was the word for "sink/drown," which comes from the same root, *t-v-a* and appears in the phrase " . . . the pick of [Pharaoh's] officers were sunk (*tuvu*) in the Red Sea."[3] I put my sister to work on this, too, for a report in her Hebrew class. We're all stumped.

Incidentally, it's not easy to find kosher food here on the Navajo reservation. Could you please send me some of those natural Israeli cookies? The sweeter, the better.

Ayyal

ADvertising Nature's ADvantages. (*Top*): Orville worked hard to produce that natural flavor; (*Bottom left*): What could be better than Natural *and* Pure? For a grand slam it carries the O-U (Orthodox Union) symbol of approval as kosher. Note that the sardines are pure despite containing calcium, iron, protein, B and D vitamins, and . . . Omega-3; (*Bottom right*): Hand Cream from Nature, 100% (. . . of what?).

5 Tevet (Dec. 18)
Netanya

Dear Ayyal,

The reason you haven't found Nature in the early Jewish sources is that it (she?) isn't there. I sent you, your rabbi, your sister, and the Navajo college librarian on a wild word chase.

Many scholars have commented on the absence of a word for "Nature" in the Bible and Talmud. The Hebrew word for Nature—*teva*—came into use relatively late, in the writings of medieval Jewish philosophers.[4] The Bible, as your rabbi pointed out, uses that root to mean "embedded" or "sunk." And the Talmud uses the word to talk about the "characteristic imprint" of something or someone. The link is understandable: A characteristic of something is indeed *embedded* in it, or intrinsic to it. From "intrinsic characteristic" the distance is not great to Nature as the "phenomenon of creation."[5]

One reason I am so interested in this subject is that I heard a fascinating lecture on the dichotomy of "Natural/Unnatural" by a chemist, Roald Hoffmann, who visited here a while ago. First he demolished effectively, or so I thought, the idea that there is a nice, clean separation between the natural and the synthetic. Then he turned around and showed how everyone, scientists included, psychologically and spiritually nevertheless prefers the natural.[6] He would be interested in our correspondence. And I like to get many people involved in whatever bothers me. May I send copies of your letters to him?

Shira

Dec. 26
Ship Rock, N.M.

Dear Shira,

Those chemists are the worst; their labs and factories are to blame for this incredible pollution we have, for cancer of the world. Maybe this one can explain why they do it. Sure, let's have him join in.

Incidentally, while I was on the chase for the first Hebrew usage of Nature (*teva*), I looked into the etymology of our English word, *N-A-T-U-R-E*. Nature and natural derive from the Latin *natura* (from *natus*, born). There are two primary groups of meanings for these words,[7] best epitomized by the oft capitalized and never capitalized variants:

> *Nature:* the totality of things in the universe (often capitalized, and—in the good old days—in supposedly gender-free English, construed as feminine)

> *nature:* the set of characteristics of an object that make it just that (object) and not another, for example, the nature of steel, or of love.[8]

I thought antonyms for "Nature" might be revealing, as they often are, about what is negated or excluded. So there is "supernatural" in the sense of "moving beyond nature, miraculous," which refers quite clearly to the first sense of Nature. "Unnatural" is another antonym, meaning "not of nature." Nature has acquired a positive valuation, "unnatural" has become burdened with negative feeling. Consider, for instance the sequence of near-synonyms: man (or woman-) made, artifactual, synthetic, artificial, unnatural. How different from each other the emotional overtones these words evoke!

So, Shira, I am ready to define my terms of the discussion. I mean "Nature" in the first sense. The world that we need to preserve.

And I'm waiting for that erotic part.

Ayyal

Jan. 19
Ithaca, NY

Dear Ayyal,

Thanks for inviting me to join in this discussion. Your definition (which Shira sent to me; in turn I will send to her copies of my letters to you) still leaves unclear what you include. Are humans and their works among the things of nature? If we include in the natural a beaver dam, why not Hoover Dam? My loafers have a label: "30% man-made material" (a euphemism for plastic). Are the leather soles not "man-made"? Signs on an Australian highway proclaim "Animals Prohibited on This Freeway." Are the prohibited animals horses, kangaroos, or reckless drivers?

Humans were clearly considered part of the same category of life as animals by the Greeks.[9] Not so in either the Jewish or Christian traditions. The covenant between God and his people is just that, a sacred pact with humans. In the Creation, humans are shaped last, and humanity specifically given dominion (or is it stewardship?) over animals.

While there are times when it makes sense to consider the biological continuity of humans and other animals, a classification that makes *all* part of Nature is self-defeating. It simply stops discussion, for nothing is then unnatural. And it is clear that wise human beings make a distinction between their actions, which are guided by a soul and reason (for some, just by reason), and actions of other creatures and the inanimate (shall we fight over that distinction too?) world. So to have a sensible discussion, a dialogue that includes environmental issues, we should make that unnatural division of distinguishing human action.

I came across two quotes from contemporary thinkers that deal with Man in opposition to Nature, in very different ways. Wendell Berry has written:

> I assume that when a man goes to the woods . . . he goes because he needs to. I think he is drawn to the wilderness much as he is

drawn to a woman; it is, in its way, his opposite. It is as far as possible unlike his home or his work or anything he will ever manufacture. For that reason he can take from it a solace . . . such as he can find nowhere else. . . . A man drawn to the woods does not go there for what has come to be known as recreation. Why should anything once created need to be re-created?[10]

An alternative view comes from Arturo Gómez-Pompa and Andrea Kaus:

> The concept of wilderness as the untouched or untamed land is mostly an urban perception, the view of people who are far removed from the natural environment they depend on for raw resources. The inhabitants of rural areas have different views of the areas that urbanites designate as wilderness. . . . Indigenous groups in the tropics, for example, do not consider the tropical forest environment to be wild; it is their home. To them the urban setting might be perceived as a wilderness. . . . In any current dialogue regarding tropical forests, for instance, the Amazon basin is usually mentioned as a vital area to be left untouched and protected. Yet, archaeological, historical, and ecological evidence increasingly shows not only a high density of human populations in the past and sites of continuous human occupation over many centuries but an intensively managed and constantly changing environment as well.[11]

To which view do you feel closer, Ayyal?

Cordially yours,
Roald Hoffmann

Feb. 4
Ship Rock, N.M.

Dear Dr. Hoffmann,

I like that first quote, of course. But the second one undermines my assumption that dividing the world into natural wilderness and man-made, sterile cities was, well . . . doin' what comes naturally. But I'm not ready to give up—I still think there is a whale of a difference between cotton and nylon. And there's a reason why cotton feels better on my body.

Respectfully,
Ayyal

P. S. Are you really a chemist? You don't write like one.

Tu B'Shvat (Feb. 5)
Netanya

Dear Ayyal,

I got your letter of Dec. 26, and the one you exchanged with Roald.

It is fitting to write about nature today, on Tu B'Shvat, the 15th day of Hebrew month of Shvat. This is the "new year" for counting the age of trees. Trees occupy a special niche in Jewish tradition. Did you know that there are laws that apply to fruit—one is that we don't eat fruit from trees in their first three years of growth. We count the years from one Tu B'Shvat to the next. That's only one of many laws limiting our harvesting (or maybe you'd call it exploiting).

Back to what we were talking about: So the word for Nature isn't natural; it came into the Hebrew vocabulary fairly late— around the 10th century. Something similar happens in other spiritual vocabularies of this world—those of painting and music, art forms that are natural for representing the beauty of Nature. If you can get to a library again, see how far back you have to go to find paintings of Nature per se; what we would call landscape.

On second thought, maybe you need a focus for your search: The root of your name, Ayyal, gives us the Hebrew words for "ram" and "deer." So look, for example, for the earliest depiction of deer gamboling, or a herd of sheep munching in a pasture, or any pastoral scene.

Shira

Feb. 18
Ship Rock, N.M.

Dear Shira,

I think you must be in conspiracy with my parents—to get me out of the corn and into books! I looked through all the art books in the public library here—hundreds of old paintings. I've never looked at art from that perspective. I see what you are driving at—there is some Nature in medieval Western painting, but it's always subjugated to the human or saintly figure. I just don't find any depictions of forests, rivers, pastures, or animals as the main subject. By chance I found a book that has illustrations with Jewish themes—graven images and all that aside. I send along a poor copy of a striking painting from a 14th-century Spanish *Haggadah* [see Plate 2]. What I think is fun is that the composition of the scene is Western. And the sheep that Moses leads are subsidiary, just as the little landscapes are in Italian Renaissance portraits. But the patterned background and the edge look like an Oriental carpet, don't they?[12]

It seems to me that it must have been the Dutch artists who first let nature be central. Look at *The Young Bull*, painted in 1647 by Paulus Potter [see Plate 3], who apparently lived to be only 29. It's a landscape, sure, and it has my namesake ram in it. But really the painting is about the triumph of nature—look how hemmed in the peasant is by tree trunks and animals. The bull is in full command.[13]

As I left the library I realized I was culturally myopic in looking at only Western painting. So I went back and found many landscapes, much, much earlier in Oriental art than in European. I enclose a poor copy of an incredible, early 11th-century painting of streams and mountains by Fan K'uan [see Plate 4].

Ayyal

Feb. 29
Ithaca

Dear Ayyal,

The reason you found no European paintings of nature before the 15th century is that religious and secular potentates—who controlled patronage of art—wanted, for their own individual reasons, the saintly or real human figure drawn as central. Dürer painted a fantastically detailed piece of turf and the head of a roebuck like none that had ever been drawn before. But those were sketches, not commissions. Nature was subjugated, pushed into the background. Given nature's direct line into the soul, it could hardly remain so forever.

There's quite a contrast between the peripheral role enforced on nature in the European Renaissance, and the Chinese representation you found. (Incidentally, I'm very happy that you were willing to go outside of our culture. . . .) One has to search in the Chinese scene for the human figure—nature appears central. But, without taking anything away from the sweep of Fan K'uan's magnificent scroll, you had better ask how authentic this nature is. Any artistic representation is selective. This particular nature is tamed and organized by the scholar/artist's gaze; you aren't looking at a natural view of nature.

I could do the same exercise in the vocabulary of music as you did in the world of painting—the sounds and musical images of nature traditionally served the purposes of other themes, religion or love. They were not allowed to be their own subject. Nature was not empowered to be by itself a musical subject.

To continue with your "pastoral" name, Ayyal, compare the two musical segments on the tape I enclose. The first is from Handel's *Messiah*.[14] The chorus sings, "All we like sheep have gone astray / We have turned every one to his own way." Listen to how "astray-ay-ay-ay" is sung. The notes literally go astray, with the soprano line rising, and the tenor falling. You can even see in the

Nature—a symbol for a religious idea. In this chorus from Handel's *Messiah* (1741), the notes stray from the straight and narrow, as we (i.e., the sheep) do.

score the sheep wandering as the music meanders [see illustration above]. Handel, born in 1685, reflects the dominant view—nature as a metaphor for religious ideas.

Some 200 years after Handel's sheep meander through his oratorio, we find ourselves again in a pasture in Richard Strauss's 1915 *Alpine* Symphony [see illustration below].[15] The selection on the tape is from the eighth section, called "The Alpine Pasture." You hear and see the tinkling of cowbells indicated in the score, and possibly the bleating of an *ayyil*, a ram, along with a yodel and a bird song.[16]

We behold nature *qua* nature: The sheep (or cows) are standing on their own four feet. They are not icons for anything else.

Best regards,
Roald

Nature—independent of religion. From "In the Pasture" section of Richard Strauss's *Alpine* Symphony (1915); herd sounds are heard.

P.S. I really am a chemist. Look in your organic textbook for something called the Woodward-Hoffmann rules. These rules are likely to wind up on your exam!

P.P.S. I just saw some really nice sheep, Ayyal—take a look at the enclosed card [see illustration below]. Even though they are in a church, they're in the context of a scene from the Pentateuch, an illustration of Moses' vocation. The grazing sheep are on a wooden panel from a fantastic 5th century C.E. door in the church of Santa Sabina on the Aventine Hill in Rome.

Moses tending sheep, in a wooden panel from the Basilica Santa Sabina, Rome.

Mar. 5
Ship Rock, N.M.

Dear Dr. Hoffmann and Shira,

There may not have been a word for nature in the Bible, but there is nature aplenty—even in the most romantic and human-exclusionary sense of the word. Desert and sea, mountain and stream, and all the creatures therein populate the Bible. Take my namesakes, the *ayyil* and the *ayyal*, the ram and the deer. They abound in some books of the Bible, as in the Psalms of David. Witness the paeans to nature in these Psalms:[17]

> *Like a hind crying for water,*
> *my soul cries for you, O God;*
>
> (Psalm 42:2)
>
> *The Lord is my shepherd;*
> *I shall not want.*
> *He maketh me to lie down in green pastures:*
> *He leadeth me beside the still waters.*
> *He restoreth my soul:*
>
> (Psalm 23:1, 2)[18]

I like the first one because it uses my name; and I know, Shira, it will not escape your eagle eyes that I've switched translations.

Ayyal

Mar. 10
Ithaca

Yes, Ayyal,

Those are great hymns to nature. Wonder, reverence—these are some of the emotions the Psalms evoke. Or to use a word with a venerable history, Nature is *sublime*, its grandeur evoking in man the sense of awe that must inevitably lead the religious person to contemplate something greater, to contemplate God.

The concept of the sublime has a lineage, from Longinus, to Burke, to Kant, to Schopenhauer. Here, I'll limit myself to some lines of Emerson:

> Standing on the bare ground,—my head bathed by the blithe air, and uplifted into infinite space,—all mean egotism vanishes. I become a transparent eye-ball; I am nothing; I see all; the currents of the Universal being circulate through me; I am part and parcel of God.[19]

Keep in mind, Ayyal, that Emerson went to Harvard nineteen years before he penned his first book, *Nature*.

Be well,
Roald

P.S. Please call me Roald.

29 Adar (Mar. 20)
Netanya

Dear Roald and Ayyal,

Roald, you miss the mark on three points in your last letter. Psalms are in no way about nature.

What is the aim of David's song? Clearly it is to bring man to God. "It is not dumb nature but Divinity that inspired the Psalmists and made them burst out into song."[20] Not only are these Psalms you cite, and all Psalms, songs of praise to the Creator, but in them (without the slightest hesitation) every wonder of nature is viewed as the product of God's creation. Or giving evidence of His might and goodness. Read the continuation of the pastoral analogy of the 23rd Psalm,

> *He leadeth me in the paths of righteousness*
> *for His name's sake.*

Yea, though I walk
through the valley of the shadow of death,
I will fear no evil:
for Thou art with me;
Thy rod and Thy staff they comfort me.[21]

Notice also that suddenly in the middle there is a switch from the distant third person "He leadeth me" to the close, intimate, second person "for Thou art with me." As one writer has said, "It is not the wonders of nature, the wonders of creation, but the wonders of the Creator, which the Psalmist admires. . . . "[22] And His benevolence, to humans.

Secondly, you confound the "sublime" and the religious. The rudiments of the idea of the sublime may have been there earlier, but the concept really developed as a *substitute* for religion in 18th-century England, an "anti-theology."[23]

Third, how can you say that nature-inspired awe "must inevitably lead . . . to contemplate God" when we know that people can view Niagara Falls and remain atheists? Emerson's stance is closer to pantheism than to seeing man's greatness in his service to the Creator.

Shira

Mar. 26
Ship Rock, N.M.

Dear Roald,

The sublime is fine and dandy, but I'm here to learn how to grow corn, some of which goes for animal and human food, and some sold for oil and syrup production. And there your gang is messing things up for my friends economically, making oil and syrup substitutes—olestra and aspartame. It's as if you want the Navajos to be stuck forever weaving blankets and making silver and turquoise bracelets.

I think that I need to understand better not so much the history of the idea of Nature, but what goes into making things that matter to me—natural oils, or sweeteners, or fibers, versus synthetic ones. So I come to the second of Shira's questions, "What is the difference chemically, between natural and synthetic?" I think I know, but I thought I would give a real chemist a chance to explain.

Ayyal

Mar. 30
Ithaca

Ah Ayyal,
You know the way to a chemistry professor's heart—ask him to give a lecture. Even if he's not a real chemist, just a theoretical one. This is a subject I'm impassioned about—why do you (we, I) eschew synthetic fibers in favor of wool in our sweaters? Reject plastic in favor of deerskin moccasins? And forgo nylon or rayon in favor of cotton socks?

Allow me to choose one of these sets of substances and deal in depth with your question. I'll pick—cotton, rayon, and nylon. At first sight, things are always simple: The first is grown in fields, the product of a living plant; the other two are products of a chemical factory.

But it's hardly simple. The domesticated cotton plant, *Gossypium*, has been bred for high yield, for better fiber, and for other advantageous properties, for hundreds of years. Modern cotton cultivars are, I suspect, genetically distant from the natural precursors first domesticated. A typical field of Egyptian cotton receives several treatments with insecticides, herbicides, and chemical fertilizers. The fiber is separated from the seed (ginned), carded, spun into a yarn. For modern shirting, cotton is also treated in a variety of chemical baths, bleached, dyed. It may be "mercerized," strengthened by treatment with lye (sodium hydroxide). Optical brighteners or flame retardants might be added. Eventually the cotton is woven into cloth, cut, and sewn into a garment. It may be blended with another fiber for strength, comfort, or some other desirable property.

That's an awful lot of manipulation by human beings and their tools, and to sharpen the point, manipulation by *chemicals*, synthetic and natural, going into your *natural* cotton shirt!

Nylon was invented in 1935 by Wallace Carothers at DuPont.[24] Its most common variant, nylon 6,6, is a polymer of hexamethylenediamine and adipic acid [see illustration below]. These are a mouthful, but actually they are just the component "monomer" molecules that are linked alternately to the long-chain "polymer" called nylon.

Where does nylon come from? To say "out of a chemical factory" has about the same level of facing up to meanings as saying that "babies come out of hospitals." The two component "monomers" which are linked up (in giant reaction vessels) to tons of polymer, derive from "feedstocks"—a generic name for the bulk materials that are the starting point for industrial chemical synthesis. The chemical origins of both monomers are in petroleum, natural gas, and the nitrogen of the atmosphere. And where does petroleum come from? Yes, yes, we know it's from an oil well. But what are the origins of oil? Natural, to be sure. This important raw material

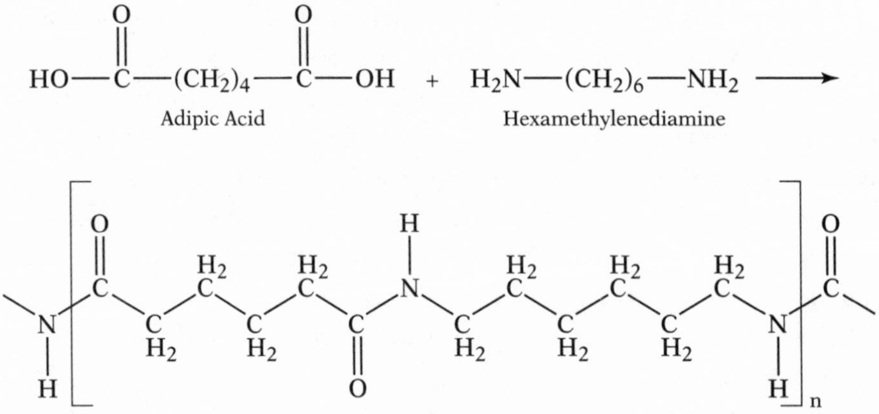

The synthesis and structure of nylon 6,6.

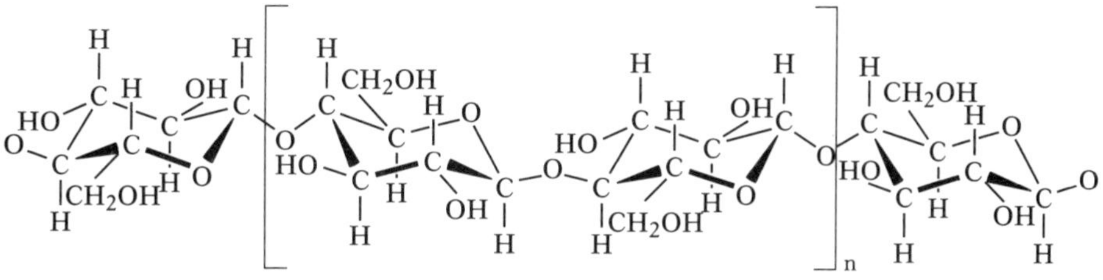

cellulose

Cellulose, the polymer in cotton and rayon.

formed from the transformation over long, long times of . . . ancient, abundant plant growth.

Rayon has a still more complicated (intellectually, not chemically) path to that lovely blouse. Rayon begins as a contemporary and most natural product, wood (actually, that wood is a product of silviculture, as much of human origin as cotton). Cellulose [see illustration above] is extracted from wood pulp and then chemically modified and regenerated. The simplest kind of rayon (it's called viscose rayon) is just regenerated cellulose, the same long-chain molecule that makes up 99% of clean cotton. I would call rayon a "semisynthetic" polymer.[25] Actually some rayon is made using cellulose from cotton, not wood!

So now that we know how these fibers we use are made, we ask the question: How much more natural is cotton than nylon or rayon? They all began life with a seed—cotton in an Egyptian field, nylon in a prehistoric jungle, rayon in a Georgia (U.S.) forest. Interestingly, it is only the seed that many million years later led to nylon, it is only *that* prehistoric seed that can lay a full claim to being natural! The other seeds were planted by human beings. And of all three plants, it is again only the nylon "plant" that can lay a full claim to growing naturally, freely, untended by humans. Then, in a sequence of human transformations, all three plants (or their remnant, the oil) eventually become the fibers of our textiles. The details differ, of course, and so does the scale. But please don't tell me that modern agriculture is all that different from a chemical factory. On some spiritual level we would all like it to be different, but it just is not. The logic of transformations (a human logic, done with human tools) is quite similar.

Let me put it in another way. Every atom of the cotton cellulose, the copolymer of nylon, the regenerated cellulose of rayon—every O, H, N, and C in these fibers—began life in a "natural" molecule. And every molecule of the cotton fiber, really no less than those of nylon or rayon, is the consequence of *human* transformations of matter.

Into these transformations go ingenuity and labor. The same ingredients that in such nicely different ways entered into the Chinese scroll you found, the drafting of the Declaration of the Rights of Man, Strauss's composing of the *Alpine* Symphony, and Maimonides's writing of *The Guide to the Perplexed*.[26]

Take care,
Roald

22 Nissan (Apr. 11)
Netanya

Dear Ayyal,

This difference you are probing with Dr. Hoffmann, between natural and synthetic fibers, is intriguing from the point of view of the *halakhah*, Jewish religious law. The fringes (*tzitzit*[27]) that you probably have on that special four-cornered garment worn by Jewish males under their shirts[28] are most likely made of nylon. And the garment itself, for reasons of cost or comfort, may be woven from a synthetic fiber. Did you know there is controversy about whether the commandment to wear the fringes applies if the garment is made from a synthetic fabric?

Fringes are worn if the garment is made of wool; linen; camel, rabbit, or goat hair; or raw or refined silk. They cannot be worn if the garment is of cow leather or deerskin. And as for nylon, rayon, cotton, or paper—the opinions differ.

If you want to delve into the reasons behind the differing rabbinic opinions, Ayyal, then I recommend the review article by Rabbi J. D. Bleich.[29] (In fact, he discusses the problems in Jewish religious observance posed by other materials not in existence during biblical and talmudic times: aluminum utensils, paper cups, disposable diapers.)

Leather is specifically exempted by the authorities from the requirement for fringes, because it is not woven. So you can see, Ayyal, that our conventional dichotomy of natural/unnatural does not help here.

One opinion cited by Rabbi Bleich advances the position that since synthetic cloth is manufactured by causing separate particles to adhere to one another, this process is comparable to weaving cloth out of individual strands of threads and would come under the rubric "woven." How does that strike you, Roald? Are strings of molecules, polymers, "woven" together? You know my chemistry is so weak. I had a course in my engineering program at Stanford, but like most engineers I resisted it. . . .

In any case, the garment in question is not worn for physical comfort—for that we have a plethora of clothing. It is worn as a physical reminder of a *spiritual concept*, of the commandment to do so, and the attendant religious value. There is no denying, Ayyal, that nature speaks to us spiritually. That leads me to the next question on our agenda: "What is the difference, spiritually now, between natural and synthetic?" Would the correspondents care to comment?

Shira

April 20
Ship Rock, N.M.

Dear Shira (and Roald),

Roald: I'm still not convinced, and I'm not going to let you get off so easily. I'll come back to you.

But first let me say to Shira that Bleich's quoted authority, may he remain unnamed, needs Roald's first-year chemistry course more than you do. Weaving could hardly be the essential disqualifier of synthetics. I had a summer job in a Hoechst Celanese plant, and so I know—all these synthetic polymers are extruded from a vat or reactor where they are "polymerized" as fine fibers through little orifices, nozzles, which in a fit of naturalizing language are called "spinnerets" (the same word as the one describing the organs through which spiders exude their silk). Then the fibers are spun like any fiber, and woven into fabric by the same mechanical Rapunzels that weave cotton and wool fabrics. Rabbi Bleich's respondent doesn't know much chemistry.[30] Nor is he likely, being a man, to have woven anything in his life. He needs to spend some time on the reservation here.

This is fun, but let me now approach your new question, Shira. I think the story is pretty clear—our spirit craves the natural, the living. You know, one of the reasons I wanted to go to Harvard is that E. O. Wilson teaches there. I came across a book of his, *Biophilia*, which makes a compelling argument for "our innate tendency to focus on life and life-like processes."

This is a spiritual matter, deep within us. And Native Americans, whom we've nearly exterminated, realized this thousands of years before us. Here is a poem entitled "Prayer" by Joseph Bruchac, who is of the Abenaki tribe:

> *Let my words*
> *be bright with animals,*
> *images the flash of a gull's wing.*
> *If we pretend*
> *that we are at the center,*
> *that moles and kingfishers,*

eels and coyotes
are at the edge of grace,
then we circle, dead moons
about a cold sun.
This morning I ask only
the blessing of the crayfish,
the beatitude of the birds;
to wear the skin of the bear
in my songs;
to work like a man with my hands.[31]

Ayyal

P.S. The cookies were great. They were natural, but they tasted like real chocolate cookies.

April 27
Ithaca

Dear Ayyal and Shira:

Shira, you ask what the difference is spiritually between the natural and the synthetic. I think the difference is *entirely* spiritual and aesthetic. And no less important for being so.

Ayyal indeed has said it well, through Joseph Bruchac's (man-made) poem. I would add that our cultural heritage (and it's also Bruchac's, who is part Czech, part Indian) leads us to favor the bucolic and pastoral, which we romanticize. Also the synthetic often comes in cost-efficient multiples, and in its design imitates the natural. While *we* are desperate for authenticity in our lives. . . . This the natural provides (even if it is shaped by us, as are the hills and fields across Cayuga Lake outside my window), in its infinite chanced variability.[32]

Incidentally, "A View Through a Window May Influence Recovery from Surgery" is the title of an article in *Science* magazine.[33] And the painted landscapes Ayyal found share certain features—trees, clearings, flowing water—across cultures.

Actually, I'm glad you quoted a poem, Ayyal, because I want to approach the question Shira posed also through a poetry of nature, though very different in time and space from Bruchac's. It is the poetry of Jews in medieval Spain.

Time marks closings and openings, sadness and joy. It is difficult to think without emotion of 1492 on the Iberian peninsula. For in the same year that Columbus, an explorer of the Spanish crown, set foot on an island in the Americas, the last Muslim kingdom of Granada was conquered by Castile and Aragon, and the Jews were expelled from Christian Spain. Al Andalus, Islamic Spain, is more than history. Over 750 years, from the fall of Visigothic Spain to Arab invaders in 711 to the surrender of Granada in 1492, Al Andalus, the western reach of the mighty crescent of Islamic rule, was a political reality, an idea, and a culture. Its loss is now tinged with irrevocable sadness.

In Iberia, a mix of Muslims (distinguishable as communities of Arabs, Berbers, and Spanish converts to Islam), Jews, and Christians (the Iberian natives who remained true to their faith, and the "Slavs," of Gothic origin) lived in relative harmony. Islam provided a well-defined, protected (if subservient) status for its monotheistic minorities. And it allowed Jews to partake, to varying degrees, of life at every level of society.

In its best times, the court life of Al Andalus provided an unparalleled setting for the intellectual exercise of poetry and philosophy. Skill in these was valued and cultivated. The same court life provided dissipation, real and ritualized. So love and lust, the joys of wine, all pursued in the garden of gardens, were celebrated in Arabic verse of immense formal subtlety. And that poetry was superbly personal and expressive.

The richness of Arabic poetry, its tropes and metrics, were emulated by many Jews, who constituted a substantial minority in Al Andalus. Jews wrote in Arabic, and then, impelled to show that their own language was capable of an equally wide range of formal devices and emotions, in Hebrew. They succeeded, and this became a great period of Hebrew literature.[34]

The remarkable courtier-rabbis (some of them) of Al Andalus wrote of wine, women, and the garden in their youth, and of repentance in their old age. And they wrote of nature in a way absent from Jewish tradition since the Psalms or the Song of Songs. Here is a poem by Solomon ibn Gabirol (born about 1020):

> *The winter writes with the ink of its rain and its*
> *   showers,*
> *With the nib of its lightning, with the hand of its clouds,*
> *A message upon the garden, of violet and purple.*
> *No human being can perform acts such as these.*
> *And when the earth becomes jealous of the skies,*
> *It embroiders its garments with flowers like the stars.*[35]

And one by Judah Halevi (b. 1075), writing of love:

> *Bear my greetings, mixed with tears,*
> *Mountains, hills—whoever hears—*
> *To ten lovely fingernails*
> *Painted with blood from my entrails;*
> *To eyes mascaraed with black dye*
> *From the pupil of my eye.*
> *Though she'll never call me dear,*
> *Maybe she'll pity me for my tear.*[36]

The setting of these Hebrew poems is much the same as that of their Arabic models—often a pleasure garden, as in this fragrant Hebrew fragment by ibn Gabirol:

> *And by the garden channels were does,*
> *hollow, pouring water,*
> *sprinkling the plants in the garden beds . . .*
> *and everything fragrant as spices,*
> *everything seemed perfumed with myrrh.*
> *   Birds were singing in the boughs,*
> *peering through the fronds of palm,*
> *and there were fresh and lovely blossoms—*
> *roses, saffron, narcissus—*
> *and each was boasting that he was the best,*
> *(though we thought every one was beautiful).*

The narcissi said, "We are so white
we rule the sun and moon and stars!"
. . . .
The buck rose up against the girls
and darkened their splendor with their own,
boasted that they were the best of all
because they resembled young rams.
*   But when the sun rose high above them all,*
I cried out "Halt! Do not cross boundaries!"[37]

Is the nature in this poem real? Hardly. This garden, as the
gardens of the Generalife in Granada, as—in another place, but not
that different a time—the exquisite temple gardens of Kyoto, these
gardens are the supreme artifice. In them are constructed canals,
fountains, and pools. In them grow imported exotic plants, the best-
manicured moss in the world, and tamed creatures. It is not at all
out of place in this beautiful garden of the mind (and yet a real
garden!) for flowers to speak.[38]

But then, in the very last line, the poet stills this exuberant
beauty by Joshua's injunction for the sun to stand still.[39] This is a
Jewish turn. Does the poet do so out of desire to make the moment
last forever? Or is it in rebuke? Maybe both interpretations have
validity.

Ross Brann, in whose course I learned of this great poetry, calls
the courtier rabbis "compunctious poets," and aptly sees in their
wonderful poems an important, continuing conflict, that

> between the desire to produce and enjoy secular Hebrew literature,
> and the awareness that the ethos of this literature, and indeed of
> literature per se, was at odds with traditional religion and
> culture.[40]

In the end, the poetry, when good, overcomes such internal
conflicts, and emerges in something new, in the service of God. As
in this poem by Moses ibn Ezra (b. 1055):

Hurry to the lovers' camp,
  Dispersed by Time, a ruin now;
Once the haunt of love's gazelles,
  Wolves' and lions' lair today.

From far away I hear Gazelle,
  From Edom's keep and Arab's cell,
Mourning the lover of her youth,
  Sounding lovely, ancient words:
"Fortify me with lovers' flasks,
  Strengthen me with sweets of love."[41]

As Raymond Scheindlin points out, the gazelle and the desert camp are standard motifs of Arabic poetry. Here they are adapted, with feeling, to Jewish religious poetry—the gazelle is sometimes God, sometimes Israel, sometimes the Messiah.[42]

In friendship,
Roald

May 5
Ship Rock, N.M.

Dear Shira and Roald,

Hey, I'm in that poem of ibn Gabirol too! Young bucks, that's me all over. And I didn't realize that the gazelle, or *ayyala* (the female counterpart of my name), is such a versatile metaphor.

I love that poetry as much as you do. And I also like the idea that these poems were written in a period of cultural assimilation. Our people could take what was of value from the high Islamic culture they lived in, yet remain observant Jews. In that way, the courtier-rabbi poets actually show me a way to live in this world.

Shira, I have not forgotten that you said there was an erotic side even to the natural/unnatural division. I've been waiting patiently, now please do tell us what that is all about!

Ayyal

23 Iyyar (May 12)
Netanya

Dear Ayyal,

I see I'm not going to be able to withhold this any longer. I think
you might be disappointed. . . .

When you visited us in Kiryat Sanz last Passover, you probably
noticed that all the married women in this hasidic enclave cover
their hair. About 98% wear wigs. The others, myself included, wear
scarves, turbans, or hats. (Actually, I have a wig but I haven't had
time to have it fixed up.) Some wear a pillbox hat on top of their
wigs as a second covering. The majority of the residents here are
Ashkenazi; but even the women in the small Sephardi minority are
starting to switch from their traditional scarves to wigs.

I have often been asked the obvious question: Why wigs? And in
the context of our present discussion, why something so artificial as
a wig, to imitate a woman's beautiful, natural hair? Well, within
Judaism, the wearing of wigs by married women evolved within the
context of three intersecting areas of religious tradition, and one of
secular custom: (1) the religious injunction for a woman to cover
her hair, (2) a general set of accepted beliefs concerning proper and
seemly behavior of women, (3) Jewish community tradition, and
(4) fashion, hardly a Jewish monopoly.

(1) The religious imperative comes from a discussion in the
Talmud, where a number of violations of precepts, biblical and
rabbinic, are enumerated whereby a wife may be divorced without
receiving her marriage settlement.[43] One example listed under the
rubric of "violating Jewish practice" is "going out with an
uncovered head." The accepted interpretation is that it is
Pentateuchally required for a married woman to cover her hair
when in a public place, and rabbinically required to cover it in less
public circumstances (her courtyard, porch, home).

A passage in the Talmud indicates the strong value attached to
hair covering. The passage speaks of a woman, Kimhit, the mother
of two priests:

> Thus their mother saw two [of her sons become] high priests on one day. . . . The Sages said unto her: "What hast thou done to merit such [glory]?" She said: "Throughout the days of my life the beams of my house have not seen the plaits of my hair." [44]

The importance of this injunction is supported by two further passages. In the ordeal of bitter waters[45] there is a fine and important detail in the Temple ceremony—the woman accused of adultery has her hair bared by a priest. This is normally the greatest indignity to a married woman. This is the Pentateuchal grounding for the Talmudic discussion of hair covering.

The second passage is a startling *midrashic* account of the wife of Onn, who in the midst of a rebellion saved her husband by strategically loosening her hair, uncovering it in the process so as to scare off his revolutionary cronies.[46]

In the Zealot fortress of Masada, which fell tragically to Roman troops in 73 C.E., four fragmentary hairnets have been found. From the evidence of the human hair found with them "they appear to have been intended to blend in appearance with the wearer's hair."[47]

(2) There is a Jewish concept of *tzniut*, of modesty and privacy in one's actions. It extends to men as well as women, but I think it is fair to say that it is applied more widely to guide (some feminists would say "to control") the life of Jewish women. From a religious point of view, *tzniut* is the cultivation of as profound as possible personal and private probity and modesty. In practice, the concept dictates that a woman need comport herself in the presence of men in such a way as not to stimulate the sexual appetites of any but her husband. This is her responsibility; the lustful men have their own sins to consider. So, as Lisa Aiken says,[48] "There is no reason that women should not look *attractive;* they are prohibited from looking *attracting.*"[49]

A married woman covering her hair satisfies the modesty laws. But those laws do not mandate a wig; that specific custom derives from another source.

(3) The customs of a community are accorded great respect in Jewish tradition, and differ from one group of Jews to another. Take

for instance Rabbi Gershom's 11th-century ban on polygyny, which applied only to the European Jewish community; Jews dwelling in Arab countries maintained biblically permitted polygyny because it did not conflict with the norms of the larger non-Jewish society where they lived. In the matter of hair covering for women both groups mandate it, but Sephardic women have until recently relied on various types of scarves and have eschewed wigs.

There are legitimate differences (to outsiders they may seem hairsplitting) among other groupings of Jews. Some ultraorthodox maintain that a wig (*sheitel* in Yiddish) does a better job at gathering in all stray wisps of hair and is in this respect more modest than a scarf, from which a lock or two may peek out. Here in Kiryat Sanz there is an interesting range. The preference of the founder of the enclave, the Hasidic Rebbe of Klausenberg, may he rest in peace, was clear: His wife, daughters, and daughters-in-law wear unique, intricately sewn turbans with elaborate bows on the right side. But most other Sanz women quietly go on wearing wigs. Many solve the problem of stray wisps by cropping their hair short, almost shaving it, upon marriage. Others add an old-fashioned pillbox hat atop their wigs. There are two sociological phenomena one can ponder here: (a) the bald fact is that the rebbe's power to enforce his understanding of correct *halakhic* practice has limits[50]; (b) a perspicacious rebbe is wise enough not to "impose on the community a hardship which the majority cannot endure."[51]

How does a Sanz woman decide which option to follow upon marriage? Two factors operate here: what the custom of her own family, especially of her mother, is, and what the preference of her groom is. A woman's desire to be attired pleasingly to her husband is given great weight by the rabbis and can override other factors.

(4) So far we discussed the hairy intricacies of the law. But now there is a historical curiosity lurking under the wig story, a story of the interplay of fashion and religion. Wigs were not originally worn by observant women in Europe. But during the Renaissance, in non-Jewish society, men and women of distinction began to wear

wigs. Wigs became fashionable. In time, Jewish women began to wear wigs, seeing it as a perfectly reasonable way to satisfy the commandment to cover their hair. In an attractive way. From Rabbi Moshe Isserles (16th century) through Rabbi Moshe Feinstein (20th century) wigs have been considered hair coverings that satisfy the injunction.[52]

Incidentally, covering one's head is not just a Jewish religious concept, for variants of the custom are found in other religions as well—note the mantillas of women and skullcaps of bishops and cardinals in the Catholic religious tradition. Why don't you go back to the art history books you looked through a while ago, and search again, this time for paintings in which women appear "bareheaded." Well into this century most women wore some kind of headgear. You'll find white caps on the Flemish women in all of Pieter Breugel's paintings. Even the 19th-century woman in a Paris street scene painted by Caillebotte[53] has a hat on when she goes out in public. Today traditional Muslim women wear at minimum scarves, and often much more, concealing the face and covering the head. Hair covering was, and in many places still is, a societal norm of feminine modesty.

How the natural and the human-made get mixed up in a wig! First in the construction, nothing to do with the religious setting: The core material of a good wig may be synthetic[54] or natural; the hair that is woven into it is often human, but usually not the wearer's. The wig certainly feigns naturalness in appearance, and is priced according to the success at just that, pretending.[55] [See Plate 5.]

The simulation that is at the core of a wig's essence is interesting in its motives, as all simulations are—Why do we make plastic tables with wood grain?

If I may let my hair down, I'd like to share with you some ponderings about the psychoerotic dynamics of wigs. The entire hair-covering topic derives from the concept that "the (head) hair of a women is *ervah*."[56] The concept of *ervah* is difficult to translate, but seems to connote something erotically attractive and stimulating. In its most technical sense the word refers to hair always hidden (by Jews and non-Jews) from public. (A parallel phrase is "a woman's voice is *ervah*."[57])

I think things that live have a special pull on us. Diane Ackerman's book, *A Natural History of the Senses,* points out that "we may breed flowers to the pitch of sense-pounding color and smell we prefer . . . but there is a special gloriousness we find only in nature at its most wild and untampered with."[58] An exquisite flower arrangement in a vase does not have the same effect on our psyche as a simple rose growing in a backyard. Similarly, a woman's hair, if cut and made into a wig, and then worn by the same woman, just doesn't have the same sensuous drawing power as the real McCoy.[59]

To be sure, *halakhah* cautions against wigs that look too real and prefers wigs that the average person (women are better at this than men) can tell are "foreign growths" (the translation of the Hebrew term for a wig).[60]

So a human-made wig that mimics natural hair (and may or may not be made from natural hair) is then worn by a Jewish woman to satisfy a religious commandment, which says to cover one's head, yet to do so while remaining natural-looking in appearance, like a woman [see Plate 6]. To be attractive, yet not too attracting.

Human beings are so nicely complicated.

Shira

May 19
Ship Rock, N.M.

Dear Shira:

> *She asks me why,*
> *I'm just a hairy guy,*
> *I'm hairy noon and night,*

Hair that's a fright.
I'm hairy high and low,
Don't ask me why, don't know.
It's not for lack of bread,
Like the Grateful Dead.
Darlin', give me a head with hair,
Long beautiful hair, Shining, gleaming, steaming, flaxen,
  waxen,
Give me down to there hair,
Shoulder length or longer,
Here, baby, there, momma, ev'rywhere, daddy, daddy.
Hair, hair, hair, hair, hair, hair, hair, hair.
Flow it, show it, long as God can grow it, my hair.
Let it fly in the breeze and get caught in the trees,
Give a home to the fleas in my hair. . . . [61]

                                              Ayyal

5 Sivan (May 26)
Netanya

Dear Ayyal,
Ah, the poetics of hair! Nearly four hundred years ago, Ben Jonson
intimated there may be erotic attraction in natural disarray:

> *Give me a looke, give me a face,*
> *That makes simplicity a grace:*
> *Robes loosely flowing, haire as free:*
> *Such sweet neglect more taketh me,*
> *Than all the adulteries of art.*
> *They strike mine eyes, but not my heart.*[62]

                                              Shira

May 26
Ithaca

Dear Shira and Ayyal:

Whew, Shira, you really had me worried when you said you were going to delve into the erotic side of nature. But you kept it decent!

Actually, I would like to tell you of another way in which people, now subcategory chemists, play with this tense edge of the natural and unnatural. This story comes straight out of the recent chemical literature.

Human beings have been put on this earth to create. Some write poems. Others write rabbinical responsa, build additions to houses, draft new civil rights legislation, dig ditches. Some make molecules—these are the chemists. All—poets, builders, lawmakers, ditchdiggers, chemists—either create something new or modify a product of nature. The God of Genesis Chapter 1 made man and woman in His image. Part of the mandate of humanity, derived from that Creation, is to create.

Is the natural different from the unnatural? Yes, on the spiritual level, as the designers of food labels know too well. No, on a material level. All stuff—natural or unnatural—is at the microscopic level molecular. And observable macroscopic properties—color, toxicity, strength, conductivity—follow from that microstructure. Synthetic molecules, carefully made, can replace natural ones. Your MSG headache is equally well induced by synthetic or natural MSG; your pneumonia cured by an antibiotic made either by a mold or in the laboratory.

Chemists do have a special way of playing with the natural. First they see it as a challenge to make any molecule nature can. And they can do it, those master builders of tiny structures. They may manage their synthesis less efficiently than nature, but then nature has had a few million more years to optimize most any process.

Second, chemists want to make molecules that aren't there in nature. Why not a molecule that looks like an icosahedron ($B_{12}H_{12}^{2-}$)? Or one like a soccer ball (C_{60})? This is real fun.

Third, they want to make molecules resembling natural ones, but better in this or that specific respect. There are polymers stronger than steel, or fats (the olestra you mentioned, Ayyal) in which you can fry your onion rings but are calorie-free, because they are not digested.

Fourth, chemists want to make molecules that are sort of like natural ones, but a little different. Why? To fool bacteria and viruses, of course. There is benefit and profit in this.

Fifth, chemists make synthetic molecules to understand nature— its highways and byways, how it got to be the beautiful way it is.

Men and women seem to find ever more ingenious ways to confound the natural/unnatural dichotomy through synthesis. Here is one example:

Nucleic acids, including DNA and RNA, are the information carriers of life. The two strands of DNA [see the illustration below], a marvelous biological polymer, are each a chain of phosphates (PO_4)

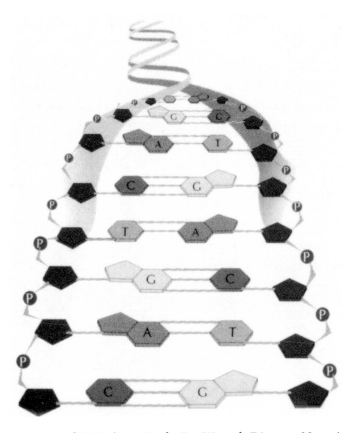

Schematic structure of DNA. (Drawing by Ian Warpole/Discover Magazine.)

alternating with sugar rings made up of five carbons and an oxygen (four of those carbons and one oxygen form those little pentagons in the model) called riboses. Coming off the rings are flat platelet-like molecular units, the "bases" adenine (A), guanine (G), cytosine (C), uracil (U, in RNA), or thymine (T, in DNA). The sequence of the bases within one strand codes the message; the pairing of the bases in that strand with specific partners in the other strand is the mechanism for replication.

There was no greater chemical achievement (even if it was accomplished by two nonchemists!) in midcentury than the recognition of this structure. And it has taken much ingenious work to explore fully the consequences and workings of what Watson and Crick had divined.

But now chemists are curious. Why this structure and not all others? Albert Eschenmoser, of the Federal Institute of Technology in Zürich, one of the deepest thinking chemists of our time, has focused on the sugars.

The sugar building blocks of DNA and RNA are pentoses, so-called because they contain five carbon atoms (chemists are prone to just occasional fits of rationality in nomenclature). Much more common than these sugars in nature (not that genetic material is rare) are the six-carbon sugars called hexoses. An example is our common glucose, maybe one of the few healthy delights left. [Ribose (a pentose) and glucose (a hexose) are shown at the top of the illustration on page 39.]

The Swiss chemist argues convincingly that hexoses are not just popular today but were more than likely to form under prebiological conditions. And he asks, "Why, then, did Nature choose pentoses and not hexoses as the sugar building blocks of nucleic acids?"[63]

Eschenmoser knew that the answer to this question was to be found not in theory but in . . . synthesis. He and his able co-workers build up an entire alternative universe—not just the sugars, but single-stranded phosphate–sugar chains. They add the bases. They do what nature chose not to do, build a "hexose–NA." [NA stands for nucleic acid; RNA and DNA are pentose–NA's. The illustration on page 39 (bottom) shows a piece of the natural pentose–NA chain and a similar piece of a synthetic hexose–NA chain.]

Top: Ribose (a pentose) and glucose (a hexose); Bottom: Pentose and hexose nucleic acids.

The alternative world in their grasp, the Swiss chemists look for differences. These are easy to find. The beauty and efficient replication properties of the pentose–NA universe (natural, ours) derives from the bases being cradled within a helix and perpendicular to it. And that in turn can be traced to the angle at which the bases B come off the sugar, relative to the chain axis formed by the phosphates and part of the pentose ring. (DNA and RNA are just polymers, long chain molecules, as were cotton, rayon, nylon.) Notice how in the hexose–NA (unnatural, even more so ours) the bases are attached at a very different angle. Computer modeling and experiment show that the hexose–NA does not form helical structures. Hexose–NA strands pair differently, pair more strongly, much less prone to the ready pairing-unpairing that is characteristic of normal pentose–DNA. That ease of pairing-unpairing is required for copying genetic material, which we like to do. The alternative universe is just not good enough, so it seems, to do what has been done.

The motivations of the chemists engaged in this work are clear. First, there is the sense of wonder. Why did nature do it this way and not another? What if we changed things just a little? The power to effect such change is in our hands. Then there is benefit and its sidekick, profit. Some of these molecules may be useful drugs, just because they are small variations on a natural theme. When curiosity and benefit find themselves on the same trail, the hands and minds of human beings seem to quicken. You can see this in the current flurry of activity in the making of slightly unnatural molecules.[64]

Be well,
Roald

15 Sivan (June 2)
Netanya

Dear Roald,

Before Ayyal gets apoplectic, let me say what I suspect is on his mind as you tell this intriguing story of "unnatural acts." What bothers many people, what bothers me, is that you are having fun, but meanwhile you are playing with the stuff of genes. In a sense, you are tampering with what God gave us. This is scary—the matter of Creation (of the essences of human beings) had best be left in God's hands.

Shira

June 10
Ship Rock, N.M.

Dear Roald,

You focus on the creation of molecules by scientists. And you attribute it to "a challenge to make any molecule nature can," to "wonder," to "profit," and to an attempt at understanding nature. You liken the work of chemists to that of any creators—"poets, builders, lawmakers, ditchdiggers"—and say that "human beings have been put on this earth to create."

I don't think that that statement is justification enough for the appropriateness of any creation. If a person's central goal is profit, or an ego-driven competition with nature, the results of that person's work will have some negative consequence. No matter if that person considers ethical issues. His considerations would be limited in value because he is already living in a world of illusions—he is separate from nature, he can improve his worth by being richer. It is the arrogance in this line of thought that can be damaging. Just as a social reformer who is more wedded to his ideology than the true concerns of the people can be harmful and the creator of a new elitism, a chemist wedded to *his* own ego can progress on a line of thought that ultimately leads to the devaluation of the human body and essence. With the power that chemical knowledge and experimentation has over all of us, this issue is of particular concern.

I will provide one more example. Poets and artists, I think, are not so much creators as expressors (and yes, it may be in a more organized form) of pure inspiration. In the *Kabbalah*, this pure inspiration is called *atzilut* and is considered the source of healing in the world. An artist burdened by his desire to be well-known will have difficulty accessing this necessary inspiration. An excess of internal structure obscures this inspiration.

So, yes, the motivations of chemists are of concern to me because of the power they hold. It is not their actions, but their paradigms of thought that are crucial to me.

In essence, a chemist who believes that it is part of our purpose to compete with nature will be able to justify taking some humans and experimenting with them. A chemist who is motivated by a purer curiosity and who lives with a respect for our universe will never think of such a mode of behavior.

Ayyal[65]

June 17
Ithaca

Dear Shira and Ayyal, answering both your letters:

Yes, some people are scared by these molecular games, as they are by genetic engineering. The range of concern is wide: Even if there are safeguards against letting loose possible new pathogens, even if these molecules save lives or improve our standard of living, what right do we have to tamper with what God gave us, as Shira says?

I do not have a full answer to these concerns, but I would begin a dialogue as follows: We *have* tampered irreversibly with nature, from the time we became a species, from the moment prehistoric man (the one recently thawed from Tyrolean ice) used a copper axe to modern times, to the ten billion chickens that now share the world with us (how many of them would be on earth if it weren't for us?). Before us, cataclysmic natural events, as destructive of other species as we are, shaped this world. We did not invent species extinctions. The difference in our transformation of nature is its scale and pace. But there is another difference, the potential for repair. Yes, I am thinking of the Hebrew phrase *tikkun olam*, repair of the world.[66]

There is a human creation (fully as unnatural as all those new molecules)—ethics—that makes every chemist think about the consequences, the possible harm, the potential ill use of the new molecules he or she makes. Science of course needs ethics. In fact, I would say that developments in modern science, for instance in "reproductive technology," have stimulated a revival in ethical thought.

We have no choice but to make these molecules, for curiosity drives us relentlessly. But we also have no choice but to worry about what we do. And act upon it. As Wendell Berry has said:

> The ability to be good is not the ability to do nothing. It is not negative or passive. It is the ability to do something well—to do good work for good reasons. In order to be good you have to know how—and this knowing is vast, complex, humble and humbling; it is of the mind and of the hands, of neither alone.[67]

Could I come back to art, which Ayyal raised? Some reasonable people in the arts think of the activity of scientists metaphorically as the killing and dissection of a free bird, the downing of the hawk circling in the sky, carving out the trajectory of its flight. Perhaps that is so for some bad science. But how different that is from the activities sketched above—from the free, inventive, intense (and responsible) play with the genetic material that shaped the bird and you and me! The way of knowing that science provides—of the making of a feather, of the bird's awesome economy of energy, of its homing mechanisms—gives us alternative ways of seeing the hawk fly. Which do not detract, in any way, from the poem, the infinite ways in which I can write a poem of that hawk. Or paint it. All knowledge enriches.

Best regards,
Roald

June 23
Ship Rock, N.M.

Dear Shira and Roald,

Enough. I want to go back to nature. In several ways. I understand that some of nature is constructed by human beings (as the gardens of Al Andalus, or the *Alpine* Symphony). And I understand that nature gains a place in our souls just through human interpretation (in words, tones, and paint) of something real, be it wild or pastoral. I hear you people—I see how attempts to separate the natural and unnatural are confounded by the experiments of chemists and great poetry.

But I still burn with anger—I can't put it any other way, Roald—about chemicals. I wouldn't mind if all of you (and I mean you chemists) worked in your labs all day and produced poetry. But you don't. You make toxic chemicals that kill, molecules that maim. Your solutions create problems.

And Shira—I can't get away from that Genesis passage: "God said unto them, Be fruitful and multiply, and replenish the earth, and subdue it: and have dominion over . . . every living thing." This passage, as well as the God-man centrality of the Psalms, leaves nature out except as poetic window dressing. It gives human beings the license to exploit nature, as long as they follow the compact between human and God. I don't like this about my own tradition. It really troubles me.

Ayyal

June 30
Ithaca

Dear Ayyal:

Anger breeds anger, Ayyal.

Religion first: There is no question that this world harbors human beings (of every religion) who do not serve their fellow creatures, who exploit them, who sell elephant tusks and vanishing species, who foul their—our—waters. Maybe, just maybe, some of them are made to feel easier about their burden of guilt by a religious framework that says the world is for man. Perhaps, but somehow I doubt it, for the notion is somewhat "philosophical." The reasons men and women choose to see personal gain as neutral, while others see in the same action sheer robbery—those self-justifications are usually lightyears away from the contemplative mode. The dark side of humans is hardly the sole property of people brought up as Jews or Christians.

Science and technology second: Yes, nature is treated as a passive being (clothed female in the old days, of course) to be poked, probed, her secrets unveiled by oh-so-clever us. Science has furthermore cultivated ethical neutrality to escape political and religious control. This not only leads to the sin of false pride, but is in the end a foolhardy maneuver. The world wars of this century, if nothing else, have taught us the folly of neutral science.

Technology, that curious hybrid of craft, science, and business, has just grown, like Topsy. A refrigerant was needed—have we got a great substance for you, an inert chlorofluorocarbon! New plastics need to be blown in the process of their manufacture—the chlorofluorocarbons (CFCs) do that so well! Shaving cream propellants, whipped cream propellants—there ain't any better ones around! Light, odorless, completely nontoxic. So CFCs, once a class of molecules of only academic interest, were made and sold, first in gram lots, then in kilotons.

As you know, these wonderfully inert (at sea level) molecules were later shown to contribute essentially to the deterioration of the fragile layer of a Manichean molecule, ozone (good up there, a bad actor in photochemical smog at sea level) in our stratosphere. Ozone filters out an important, harmful (to us) part of the sun's predominantly benevolent, much-desired radiation. There was a cursory consideration of the harmful effects of CFCs, and there seemed to be none. They were nontoxic to humans, animals, and plants. They just floated up, up, and away. And up there, there are no people. So nothing to worry about. . . .

So we've learned something—to worry about carcinogenic smoke, to worry about drugs (such as thalidomide) that cause birth malformations, to worry about watching seemingly harmless gases that down here fulfill desired functions. But there always seems to be a new danger that escapes our ever-more-worrying vigilance, doesn't there?

It is clear that our metier is to push through from discovery to megaton production the making of entirely new materials in a time span of around twenty years. The earth, Gaia, has restoring forces, but such a wholesale attack by our immense creative intelligence on these restoring forces, in the geological equivalent of the blink of an eye, tests those forces severely. The pessimistic outlook is that the Earth will survive, have no fear. Just an earth without us.

I will respond, Ayyal, but I have to indulge myself in one spot of nastiness. This is to confront the romantic paganism of some environmental proponents (yes, you too, Ayyal), reaching for an imagined past that is only in their minds. Here is what one of the most important and beautifully written early condemnations of the Judeo-Christian/science role in the environmental crisis has to say:

Christianity, in absolute contrast to ancient paganism and Asia's religions (except perhaps Zoroastrianism) not only established a dualism of man and nature but also insisted that it is God's will that man exploit nature for his proper ends.

At the level of the common people this worked out in an interesting way. In Antiquity every tree, every spring, every stream, every hill had its own *genius loci,* its guardian spirit. These spirits were accessible to men, but they were very unlike men; centaurs, fauns, and mermaids show their ambivalence. Before one cut a tree, mined a mountain, or dammed a brook, it was important to placate the spirit in charge of that particular situation, and to keep it placated. By destroying pagan animism, Christianity made it possible to exploit nature in a mood of indifference to the feelings of natural objects.[68]

It seems like it was a nice world. But in the real world, before there was Christianity or Judaism, or today in areas of the world where these religions do not dominate, where animism and Buddhism reign, for instance, there is little sign that either interhuman ethics, or human-nature interactions are any "better." Slaughter of men was not impeded by the absence of Christianity and Judaism in the India of partition times, in Timor, in Rwanda and Burundi. And I think Native Americans did not think of preserving buffalo as they hunted them. They just moved on to the next herd. They were lucky there were not too many hunters. . . .

Reaching into a dreamy, constructed past is just a way to avoid the present, where difficult ethical choices need to be made.[69]

Now some of the anger is out, and so let me respond more calmly: Knowledge is the basis of action, in moral or material matters. There is a place where the fire engine needs to be sent. Without knowledge, and without faith, action is at best random, of little effect in either the moral or material realm. Science is not neutral. But it does provide the knowledge on the basis of which political (and that is meant positively) human beings may act.

It was science, not anything else, that provided knowledge of the mechanism of action of the CFCs on the ozone layer. This was accomplished on the basis of laboratory experiments and theory by Harold Johnston and his co-workers (initially worrying about the effects of a commercial fleet of supersonic transports in the stratosphere), and by F. Sherwood Rowland and Mario Molina. Their analyses, suspicions, and conclusions were published prior to the discovery of any depletion of the ozone layer, itself a finding of great skill by atmospheric chemists.

Thus some scientists believed early on that unbridled use of CFCs could damage the ozone layer. Other scientists disagreed. And, as you might expect, the makers of the gases put together superbly qualified "defense teams" to find fault in the arguments of the scientists whose findings threatened the livelihood of these companies. A scientific dispute was transformed into a negotiated political one.

The problem arose from the making of a group of molecules, and the finding of a utility for them. But there was no hint of a problem with the ozone layer, nor any movement to a solution, without science.

The environment *is* in danger; our profligate population growth (bigger in those areas of the world, incidentally, where the Judeo-Christian ethic is *not* dominant—I'm still sore) has put immense pressures on our agriculture. Our communication skills, those TV transmissions ranging the earth, have rendered everyone's desires commensurate—it is impossible for people to remain ignorant of a better life, one nice old mechanism for keeping lots of people happy. The pressure to produce, in bulk, is great. And technology, coupled, yes, with human greed, has hurt many.

Scientists have a great responsibility to the environment. If not we, who then . . . will know. There are signs that we are finally taking that responsibility seriously.

In friendship,
Roald

July 3
Ship Rock

Dear Roald,

You've made some interesting points . . . some. And about the buffalo, you may be right. It turns out that the issue of man's dominion comes up in a Cheyenne legend that humans won a race against the buffalo and their allies (elk, antelope, and deer—in the *ayyal* family) and thus man gained the right to use buffalo and those allies for food, shelter, and clothing. That is the part of nature that Cheyenne culture needed, and man gained dominion, the entitlement to hunt and kill.[70]

So I am willing to concede that religiously sanctioned "dominion" is not a negative thing . . . as long as it isn't overdone.

Ayyal

17 Tammuz (July 4)
Netanya

Dear Ayyal,

I want to go back to those killing chemicals and maiming molecules of yours. We have been coexisting with them for a long time. The Mishnah—we're talking about over 2000 years ago— demonstrates concern that the natural not be despoiled nor polluted. Two examples of many: A city should be surrounded by a green belt that cannot be rezoned[71]; tanneries (with their foul odors) must be kept 50 cubits (about 25 meters) from the nearest dwellings.[72,73] But Jewish religious law recognizes that often there are hard choices before us, and room must be made for economic productivity.

Dyeing of textiles is of particular interest to Roald and to me. In the past, dyeing polluted in much the same way that tanneries do. Here is a responsum written in the 16th century, by Rabbi Shlomo Cohen of Greece:

The damage caused to the townspeople by the vats used by the dyeing industry is extremely great and has to be considered as similar to that of smoke and bad odors. However, since the textile industry is the main basis for the livelihood of the people of this town, it is incumbent on the neighbors to suffer the damage. This is an enlargement of the principle that where a person is doing work that is essential to his livelihood and which it is not possible to do elsewhere, the neighbors do not have the right to prevent it.[74]

The trade-off between environmental quality and livelihood persists today. Nevertheless, perhaps young people such as you, Ayyal, will find alternatives so that one won't be at the expense of the other. It's hard to do this without science.

Shira

P.S. If you can get access to the Internet out there, you might check out the Web site of the Israel Union for Environmental Defense at http://www.serve.com/IUED.

July 14
Ship Rock, N.M.

Dear Shira,

You sound worse than Roald in his most patronizing moments (sometimes he's OK!). A technological fix is not what we need. Or at best it will be a patch, and our seemingly infinite capacity for fouling our nest will always be ahead of *tikkun*.

I think we need a fundamental change in outlook. I may have given you the impression that I'm just an eco-hippie, romanticizing the Native American communion with nature. I'm not, Shira. I see the difficulties of Navajo life (and this is a successful tribe!), what drives some of my friends to drink and depression. Though for that I lay a good measure of the blame on the aggressive, eating-up mentality of what passes as American culture, and which these people must confront.

But I have not given up my learning, in the Jewish sense,
Shira. I think there must be a transformation in us and it must be
spiritual, for the soul guides the body. How impoverished,
materialistic is science, thinking otherwise!

Recently I found some encouragement and guidance in a
comment made by Abraham ibn Ezra in the 12th century. He
is writing on Psalm 115, you know the one that speaks of the idols
of the nations:

> *They have mouths, but they speak not:*
> *eyes have they, but they see not:*

Later on, the Psalm becomes a benediction:

> *Ye are blessed of the Lord*
> *which made heaven and earth.*
> *The heaven, even the heavens, are the Lord's:*
> *but the earth hath He given to the children of men.*

Ibn Ezra writes of this last passage:

> The ignorant have compared man's rule over the earth with God's
> rule over the heavens. This is not right, for God rules over
> everything. The meaning of "He hath given over to man" is that
> man is God's steward over the earth. . . . [75]

This is how I feel, Shira and Roald. The Arabic word that ibn
Ezra used was *paqid*—I think such a man was an official entrusted
with special responsibility for a task. I like that.

Ayyal

July 26
Ithaca

Dear Ayyal:

I also like that ideal of stewardship, and I think I can understand your notion, with which I agree, in another dimension. I send along separately a remarkable book that I recently read. It's *The Lonely Man of Faith* by a great Jewish thinker who passed away a few years ago, Rabbi Joseph Soloveitchik.[76] I see in it a lesson on how we might respond to your concerns, which are also mine.

Everyone who has read Genesis will have noted that there seem to be two Creations conflated into one. In Genesis 1 it says:

> And God said, "Let us make man in our image, after our likeness. They shall rule the fish of the sea, the birds of the sky, the cattle, the whole earth, and all the creeping things that creep on earth." And God created man in His image, in the image of God He created him; male and female He created them. God blessed them and said to them, "Be fruitful and multiply, and replenish the earth, and subdue it. . . ."

In Genesis 2 God forms man again:

> When the Lord God made earth and heaven—when no shrub of the field was yet on earth and no grasses of the field had yet sprouted, because the Lord God had not sent rain upon the earth and there was no man to cultivate the soil, but a flow would well up from below the ground and water the whole surface of the earth—the Lord God formed man from the dust of the earth. He blew into his nostrils the breath of life, and man became a living being. . . . The Lord God took the man and placed him in the Garden of Eden, to cultivate it and keep it.

This latter man is alone, and needs a helpmeet. You know how it goes on from there. The story of the two Adams has puzzled many. Literary scholars have seen in the two Adams an imperfect merger of two texts.

Rabbi Soloveitchik sees something deeper—two sides of humanity. In a beautiful piece of writing that I cannot do justice to except by quotation, he characterizes the two Adams:

> Adam the first who was fashioned in the image of God was blessed with great drive for creative activity and immeasurable resources for the realization of this goal, the most outstanding of which is his intelligence, the human mind, capable of confronting the outside world and inquiring into its complex workings. . . . God, in imparting the blessing to Adam the first and giving him the mandate to subdue nature, directed Adam's attention to the functional and practical aspects of his intellect through which man is able to gain control of nature.

Adam the first, Soloveitchik says, wants to be human, to discover his identity. And he acquires "dignity through glory, through his *majestic* posture vis-à-vis his environment." The Rav (the honorific by which Soloveitchik was generally addressed) relates the personality of Adam the first to that of the modern scientist, the transformer of nature who does not necessarily understand it.

Soloveitchik describes Adam the second as a very different creature. Alone, so alone at first, he is led to think, to ponder the questions of Why? What? Who? To answer these questions, Adam the second

> does not create a world of his own. Instead, he wants to understand the living, "given" world into which he has been cast. Therefore he does not mathematize phenomena or conceptualize things. He encounters the universe in all its colorfulness, splendor, and grandeur, and studies it with the naivete, awe, and admiration of the child who seeks the unusual and wonderful in every ordinary thing and event. While Adam the first is dynamic and creative, transforming sensory data into thought constructs, Adam the second is receptive and beholds the world in its original dimensions. He looks for the image of God not in the mathematical formula or the natural relational law but in every beam of light, in every bud and blossom, in the morning breeze and the stillness of a starlit evening.

Soloveitchik continues:

> Adam the first He told to exercise mastery and to "replenish the earth, and subdue it"; Adam the second, to serve. He was placed in the Garden of Eden "to cultivate it and to keep it."*

Too many of the critics of the alleged Judeo-Christian responsibility for the environmental crisis have chosen to castigate Adam the first. Some scientists, in a fit of 19th-century confidence, have chosen to emulate the same Adam. It is time to see both Adams, irrevocably choosing to live, not without anguish, but to live, on the only earth given to them. To us.[77]

To life,
Roald

Aug. 25, summer's end
Ship Rock, N.M.

Dear Roald and Shira,

I identify with your discussion of the two Adams within us, Roald. Or, as I see them, the Adams of order and of chaos, of ego and of id, of masculinity and of femininity.

I grew up in a way that was much more in line with the first Adam. To me, logical reasoning was everything. Intelligence and cleverness were two traits that I had and which I treasured highly. . . . Through time and experience, I began to understand the limitations of my world of reasoning; it lacked relevance.

In essence, I believe that the first Adam is only valuable as a servant to the second Adam. Structures are only as good as the meaning they transmit. The excess of structure and ego is what distances us from each other and from God.

* We use various translations for the Genesis citations; our translations differ from the ones used by Rav Soloveitchik in his essay.

I want to strengthen the "second Adam" in me, not because I rejected the first, but because I want to utilize him to be more of a benefit to myself and to other sentient beings.

The corn is in, it's time to think about going back to school. I'm glad, Shira, that you got me started thinking about what nature really is—though I had been through many of these ideas in my own mind. I'm also glad, Roald, even if you are a chemist, that you made me see that deciding what is natural or not is not so simple. And pointed me to some great poetry and Rav Soloveitchik.

Let me go back to that Bruchac poem, which means a lot to me. When Joseph Bruchac speaks of his desire to see himself as a part of the family of living beings, he is confronting an aspect of his own struggle to be closer to God. The trees, animals, and the sea are one part of this universe and of this struggle. So, in their interest and in ours, I hope that all people will one day focus on developing their love for the things in this universe that were not created by us.

Ayyal[78]

A Sukkah *from an Elephant*

Moritz Oppenheim's *Sukkah*

Starting in the 1860s, the German-Jewish painter Moritz Oppenheim began producing a set of twenty "pictures of traditional Jewish family life" that brought him considerable fame in his lifetime. The book version of the series may have been the most popular Jewish book ever published in Germany.[1] One of the scenes features a *sukkah*, a booth constructed by Jews for the autumn festival of Sukkot, at times called the Festival of Tabernacles.

A well-to-do Jewish family, perhaps from Frankfurt-am-Main, is seated at the holiday table in the *sukkah*, erected in the leaf-strewn yard. We peer in (see the illustration on page 56) through the curtained doors to see the family patriarch making the traditional holiday blessing over what is probably homemade raisin wine, while the *hallah* loaves are covered with delicate linen damask. His wife, holding the baby, sits at the table with the other family members. As the buxom maid brings the steaming chicken soup, the family cat watches her, hoping that some will spill from the porcelain tureen. Two German schoolboys peek in at the curious scene and probably wonder: Why on earth are these Jews eating outdoors in a weird booth on a chilly autumn day?

The booth is an integral part, the religious essence, of Sukkot. And that booth is a human construction, full of a very special play of the natural and the artifactual. The *sukkah* serves us in this chapter as a way in to the sources of Jewish tradition. And it enriches our discussion of the natural/unnatural dichotomy that pervades science and religion.

Moritz Oppenheim, *Sukkot*, 1885. Oscar & Regina Gruss Collection.

An Autumn Festival in Ancient Israel

God spoke to Moses, instructing him to say to the Israelites as follows:

> Mark, on the fifteenth day of the seventh month, when you have
> gathered in the yield of your land, you shall observe the festival of
> the Lord [to last] seven days: a complete rest on the first day, and a
> complete rest on the eighth day. On the first day you shall take the
> product of the *hadar*[2] trees, branches of palm trees, boughs of leafy
> trees, and willows of the brook, and you shall rejoice before the
> Lord your God seven days. . . . You shall live in booths seven days;
> all citizens in Israel shall live in booths, in order that future genera-
> tions may know that I made the Israelite people live in booths when
> I brought them out of the land of Egypt, I the Lord your God.[3]

The injunction is to celebrate a harvest festival, a common occasion in many cultures. The booths, of which Moritz Oppenheim portrays a genteel German-Jewish version, are a particularly Jewish feature.

. . . and Wartime Kovno

In 1941, some thousands of years after Moses received this command, in the town of Kovno in what is now Lithuania but then was the German-occupied Soviet Union, Rabbi Ephraim Oshry sits down in hiding to write the following response to a question. Around him rages the destruction of his people. The Jews of Kovno are confined to a ghetto, only months separating 99 percent of them from death at the hands of the people whose civilization made the Oppenheim paintings possible.

Question: With the approach of the holiday of Sukkot, ghetto prisoners set up a *sukkah* in a hidden alleyway made from boards taken from workshops where Jews worked as slave laborers for the Nazis and cut down to size. . . . Is it permissible to fulfill the *mitzvah*[4] of *sukkah* with raw materials that are stolen?

Response: In order to respond to this query, we must refer to several ancient source texts. The Talmud, tractate Sukkah, page 30a, discusses the problem of a stolen *sukkah* in the context of the prohibition against doing a *mitzvah* that involves a transgression. In the case of a stolen structure, the monetary equivalent must be paid to the owner. Another concept relevant here is that if the owner of a lost or stolen object gives up hope of it ever being returned, then ownership is in a sense relinquished. In all probability, the lumber in question had been stolen by the Nazis from the Lithuanian Jews who had been involved in the lumber business in the area around Kovno. The Jewish owners had given up hope since they knew they were slated for annihilation. Therefore whoever took the lumber from the Germans to make the *sukkah* took it after it had left the possession of the original owner who had surely forgotten that the lumber ever existed. In the unlikely case that it did belong to the Nazis, the latter were in the category of "pursuer" (*rodef* in Hebrew), that is, one who pursues another to kill him.[5] Therefore I rule that it is permissible to take the lumber in the first place.

In a reflection after the war, Rabbi Oshry added:

> Thank God many Jews in the ghetto fulfilled the *mitzvah* of
> *sukkah*. The Nazis never did realize that the Jews they were tram-
> pling one moment were fulfilling the commandment of *sukkah* the
> next moment. While the Germans carried out their *Estland Akzion*,
> a Jew known as Reb Zalman der Blinder hid in this *sukkah* and
> was saved. The Germans looked through all the barracks, but did
> not look for Jews inside the *sukkah!*[6]

Imagine Rabbi Oshry fielding questions like this one in the
midst of the Holocaust. (Oshry's wartime responsa are collected in
five volumes!) Tragic as the circumstances are which led to the questions, the responsa may be seen in a different light. They are
testimony to the human dignity and belief in God of a people—for
even there, even *then*, "Out of the Depths" (the Hebrew title of his
responsa), religious Jews remained concerned that what had to be
done surreptitiously and in constant fear be done in a ritually
appropriate way.

It behooves us, whether simply for reasons of curiosity in this
tenacious faith, or in memory of those who asked the questions,
cared about the answers, and perished, to see what it is that moved
them, what shaped the mind of the people, what took them from
Leviticus to Frankfurt and Kovno. And past that. The important
concept in this religious tradition is the term we explained in the
prologue, the *halakhah,* or "the way."

The Dual Torah

The universe of Jewish religious thought has at its twin hubs written
and oral law. The former is familiar to the non-Jewish public as
Scripture, enough so that it is often known by heart and oft-quoted.
This written law, what is called by Christians the Old Testament or
Holy Writ, consists of the Torah (the five books of Moses or the
Pentateuch), the Prophets, and the Writings. The Torah[7] is still written today on a scroll, as it was centuries before the Common Era
(c.e.). (A segment of such a Torah scroll is shown in the illustration
on page 59.)

The Jewish people do not live according to the Bible. Not that
they disrespect the Book, but it hardly suffices. It would be impossible to answer all the questions concerning how one should conduct

A Torah scroll, open near the passage commanding the observance of the Sukkot festival. (Photograph by Kenneth L. Fischer.)

one's life vis-à-vis food, labor, rest, holidays, clothing, marital life, social obligations, political structures, legal issues, finances, and so on, based solely on the verses of the Torah or even on all twenty-four books of the Bible. Interpretation and an evolving comprehensive legal structure are needed. This was clear even in ancient Israel, and from that necessity the oral law evolved.

The oral law is a vast, unending tradition amplifying, explaining, elucidating, and applying the precepts of the written law. According to traditional Jewish theology both the oral and written law were revealed at Sinai and in the ensuing centuries the oral law was redacted, written down, and codified in different forms for a variety of purposes. At the same time it is clear that parts of the oral law had definite identifiable human authorship (and there is a theological justification to oral law "not being in heaven," to which we will return in Chapter 3). And the oral law was formed in well-defined, historical circumstances (in the land of Israel, in Babylon, in medieval Troyes, in Kovno, and elsewhere).[8]

The oral law consists of the living line of Talmud, *midrash*, commentaries, codes, and responsa. Let us describe these sources one by one, and then show what they add to the biblical injunction to celebrate the festival of booths.

The Talmud   Venerated and studied by Jews, burned in intolerant times by their oppressors, the Talmud is a Britannica-size document of some complexity. The word "Talmud" means "what is told, what is learned." The Babylonian Talmud was put in its final form around 500 C.E.; the less studied Jerusalem Talmud had been redacted about two centuries earlier.

The contents of its sixty-odd tractates are the way of the world—they include the time of day when the morning prayer need be recited, when a man and his wife should have relations, and how one decides whether a side of beef found in an alley is kosher (ritually suitable) or not. The Talmud's contents shape (but do not entirely define) "the way"—the *halakhah*.

This brings us to two genres in the Talmud central to our discussion—*halakhah* and *aggadah*. *Halakhah* is Jewish religious law. The word has a peripatetic root, meaning "to walk." *Aggadah* is the nonlegal portion of the oral law—it consists of anecdotes, proverbs, homilies, and just plain good stories.[9] Not at all devoid of ethical import, and often full of the psychological insight that makes the oral law such a deeply human document, the Talmud is mostly *halakhic* in nature, but in fact most of what we will quote will be from the *aggadic* material.

Midrash   A *midrash* is a statement that uses a biblical verse as a springboard. The statement can be in the form of a vignette, question, law, homily, legend, pun, or fable. *Midrash*, exegesis, is ancient material that is in fact contemporary with Talmud, and sometimes even precedes it.

The *midrash* contains some *halakhah* (featuring almost the same cast of rabbinical sages engaged in serious discussion as the Talmud does). But there is more *aggadah* in the *midrash* as a whole than in the Talmud. This has brought about a (misleading) conception, prevalent in the lay Jewish community, that *midrash* is just storytelling or legend. It is much more than that.

Commentaries   The centuries following the redaction of the Talmud were hardly devoid of intellectual activity. There is a religious injunction to interpret the Torah in every age—with respect to be sure, but still to interpret it. How this is done by those who aspire to add to the achievements of revered predecessors (and how this relates to problems of authority in science) we will see in later chapters.

Medieval Europe and the far reach of Islamic rule witnessed roughly at the same time a special flowering—nay, an explosion—of brilliant religious thinking. We will have occasion to cite repeatedly the commentaries and texts of Rabbi Shlomo ben Yitzhak (acronym: Rashi) of Troyes (1040–1105) and Nahmanides (Rabbi Moses ben Nahman, acronym: Ramban) of thirteenth-century Gerona in Aragon. These men differed in temperament and philosophy but their authoritative and perceptive contributions were recognized early, and the Bible and Talmud as printed today, by convention include their astute comments.

Codes   The need to specify laws succinctly without being diverted into intriguing side discussions was met by scholars whose mastery of the corpus of talmudic literature enabled them to organize and codify the laws. The codes represent a trade-off: They are lucid and succinct, but don't explain the "why?" Talmudic intellectual debate and discussion is given up for the codes' clarity and conciseness. The two most cited codes are those of Maimonides (Rabbi Moses ben Maimon, acronym: Rambam, 1135–1204, Cordoba and Egypt) who wrote the *Mishneh Torah,* and of Joseph Caro (1488–1575, Safed, Galilee) who wrote the *Shulhan Arukh.*

Responsa   A word of Latin origin, responsa (sing. responsum) refers to the answers of a religious authority, a learned rabbi, to a query. As questions arose and scholars responded to them, a body of written questions and answers accumulated, beginning about the sixth century C.E. For example, a Jew from Baghdad might send an inquiry to Rabbi Shlomo ben Aderet who lived in Barcelona around 1250. The rabbi would write an answer and his students would copy it down (no photocopiers then). He would send the answer to Baghdad and retain the copy in manuscript. Two-and-a-half centuries later, with the arrival of Gutenberg, they were printed. We have 7,000 responsa from this one rabbi! The questions and answers of thousands of such scholars have become a public asset; there are some 250,000 extant responsa. Responsa are still being written today, vivid testimonials to the ability of ancient law to cope with modernity.[10] (In subsequent chapters we will also look at some of the problems tackled by contemporary responsa.) Who or what confers authority on a rabbi, so that his responsa are universally respected? What happens when there are conflicting responsa? These are questions fraught with inherent complexity.

The responsa of Rabbi Oshry during World War II cite every component of this body of Jewish religious tradition—Bible, Talmud, *midrash,* codes, commentaries, and other scholars' responsa— to make agonized religious decisions in desperate times. It was never very different.

An Elephant and the Material Science of a *Sukkah*

Each of the components of the oral tradition contributes to answering some of the complex questions that arise from the skimpy Biblical command: "... you shall live in booths. . . ." How many walls? What shape? What height, length, width? Which materials? Who should live in them? What if it rains? Some answers are given in these examples from the various constituents of the oral law.

Talmud To know how many two-by-fours you need order from your lumber company we must turn to the talmudic portion of the oral law. The Talmud has two important parts—the Mishnah (composed approximately 30 B.C.E.–200 C.E.) and the *Gemara* (200 C.E.–500 C.E.). These record often contentious debate among generations of religious scholars. Because of the oral nature of the initial transmission, the earlier text, the Mishnah, is often contracted, formulaic, and structured in certain ways, which lend themselves to memorization. The text is also at times cryptic, demanding deep study. The *Gemara* is a more narrative record of discussions on the Mishnah in Babylonian or Palestinian academies.

The discussions are laid out on a page of printed Talmud to enable an incredible intergenerational conversation. An entire page of such a discussion is shown in the illustration on page 63; the semiotics of the page enable you to listen, and sometimes participate, in the back-and-forth interplay of ideas, argument, counterargument . . . all for the sake of Heaven but not in heaven. To penetrate into its depth, to enter its cognitive world with its unique logic, the Talmud must be studied with a teacher. The languages of the Talmud are Hebrew and the vernacular of the time, ancient Aramaic; there is almost no punctuation and no differentiation between upper- and lower-case letters[11] so that it is difficult to know when a sentence begins.

The illustration on page 63 displays the different elements on a page of Talmud. A sentence or two of Mishnah is followed by the

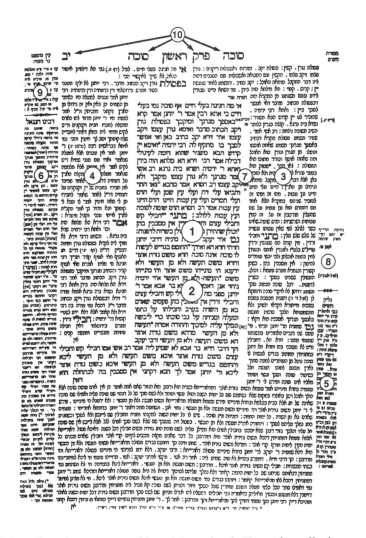

A page of the Talmud, Tractate Sukkot. (1) Mishnah ("As side walls, however, all of these may be used."); (2) *Gemara;* (3) Rashi (d. 1105); (4) Tosafists (1100–1300, Provence); (5) glosses of R. Akiva Eiger (d. 1837); (6) commentary of Rabbenu Hananel (d. 1055); (7) references to biblical sources; (8) cross-references to related issues in Talmud; (9) cross-references to classic *halakhic* codes; (10) page locators (folio number 12, tractate name, chapter number I, and title, "Sukkah").

Gemara discussion. You get caught in the crossfire between Rashi (his commentary printed in the inside margin of a page, safe from wear and tear) and his grandsons (the Tosafists, found on outer margin). Numerous other commentaries, as well as cross-references (similar to today's footnotes and references), are arranged in the margins surrounding the main text.

Let's move to the material at hand, the material allowed for construction of a *sukkah* and its configuration. This is nontrivial stuff—the Babylonian Talmud has a whole tractate of fifty-six two-sided pages called Sukkah!

The Mishnah records debates among the rabbis over the requirements for a *sukkah*, some of the controversies being resolved and others not. In bridging the gap between the written and oral law, clues are adduced from the Torah passages about the festival in Leviticus and a passage in Deuteronomy 16:13, which instructs "After the ingathering from your threshing floor and your vat, you shall hold the Festival of Booths for seven days." The interpretation is that the roofing should have the characteristics of threshing floor gleanings.

Finally, the Mishnah outlines some basic requirements for a *sukkah* that is *halakhically* valid (Hebrew: *kasher*, kosher). It raises myriad possibilities which we probably never would have thought of, as it tries to poke around the border of acceptability to find the limits of the religious notion: A double-decker *sukkah?* One on a camel or a ship? With a growing tree as a roof? As a wall? One formed by a four-poster bed? Dug out of a haystack? Cone-shaped? (See illustration on page 65.)

The Mishnah stipulates the number of walls for a *sukkah* as three, their minimal (10 *tefahim*, today about 38 inches) and maximal (38 feet) height, and minimal area (2 feet square). Here, for example, is the prescription for the roof material:

> If he trained vines . . . over [a *sukkah*] . . . it's invalid. If he cut them, it's valid. This is the rule . . . everything not subject to defilement and growing from the ground may be used [as a roof for the *sukkah*]. Tied bundles of straw, of wood, and of twigs must not be used to cover the *sukkah*. All of these become kosher, however, if the bundles are loosened. As side walls, however, all of these may be used.[12]

The Mishnah establishes the principle that everything can be used for a *wall:* bundles of straw, wood, twigs, and by implication, bricks, stone, glass, cloth, and metal. There is even one material that we suspect the reader wouldn't dream of, but the rabbis of the *Gemara* propose in their no-holds-barred exploration of the problem.

Efrat and Michael Leibowitz, in center, decorating a *sukkah* in Israel, surrounded by sketches the then twelve-year-old Sima Morell made when she completed studying the mishnaic tractate Sukkah. One of the configurations of a *sukkah* shown is not valid. (Photograph by Kenneth L. Fischer.)

What about using an animal for a wall of a *sukkah*? An elephant would have the requisite height (more than 38 inches and less than 38 feet), so why not?

Here is the *Gemara* section of the Talmud that follows the above-cited Mishnah:

If the animal were used as a wall for the *sukkah?* R. Meir declares it invalid, and R. Yehudah declares it kosher, for R. Meir used to say: Whatever contains the breath of life must not be used as a wall for the *sukkah,* nor as side-beam to an entry, nor as an enclosure to a well, nor as a covering of a grave; and in the name of R. Yosi the Galilean it was said: Also a writ of divorce must not be written on it. What is the reason of R. Meir? Abaye said: because it may die. R. Zera said: because it can escape. . . . If an elephant were tied to a wall and used as a wall, [all agree it is kosher], because even if it should die, its carcass measures more than ten handbreadths high. In what they differ is, when the elephant is not tied: according to those who fear its death, it is valid; but according to those who fear its flight, it is not valid. . . . But is there not an open space between the animal's legs? [It refers to] where he filled it in with branches of palm and bay-trees. But might it not lie down? It refers to where it was tied with cords from above. . . . Let us see: according to both sages—who apprehend death or flight—biblically it is valid as a wall; and only as a rabbinical precautionary measure is it forbidden.[13]

The relevant segment of the *Gemara's* discussion is not easy to follow—for instance, what do *sukkah* walls, well enclosures, grave covers, and divorce documents have in common? Well, in the spirit of everything in the world being connected to everything else, an animal could be used for all these disparate items! The discussion is sometimes tightly logical, sometimes digressive. These are real people engaged in down-to-earth discourse of momentous matter, much as a physicist tries to ascertain the limits of a model.

In essence four objections are raised, followed by retorts. What if the elephant wants to escape? So put it on a leash. What if it dies? Even if jumbo shrinks a bit, the carcass still meets the minimal wall-size specs. What about the space between its legs forming a hole in the "wall"? Fill it with palm branches. What if it sits down? Tie it with cords from above. The bottom line is: Yes! Rabbi Meir's opinion is cited, a sign of respect, but the ruling is according to Rabbi Yehudah—an elephant, if you can find one, is ritually fit to serve as a *sukkah* wall . . . but not to eat.[14]

Most certainly the talmudic sages were not besieged by congregants with such kosher elephants (see illustration on page 67). Rather, the discussion is a search for boundaries; the sages add flesh and blood to the bare bones of the Mishnaic definition.

The discourse in the Talmud has a living feel, at times an edge to it. Here is a snippet, dealing with the question of the strength of materials used (measured by the *sukkah's* ability to stand in an average wind). In this *aggadah*, from the very same page of Talmud, Rabban Gamaliel clinches a debate using a terse put-down of Rabbi Akiva:

> If a *sukkah* were made on a ship, Rabban Gamaliel rules it invalid and Rabbi Akiva rules it valid. It happened once that Rabban Gamaliel and R. Akiva were on a ship and R. Akiva constructed a *sukkah* on the ship. On the morrow a wind blew it off and Rabban Gamaliel said to him: Akiva [so now] where is your *sukkah*? [15]

Midrash Setting out from a biblical verse, "This is my God and I will glorify Him,"[16] a *midrash* touches several bases:

> Rabbi Ishmael says: And is it possible for a man of flesh and blood to add glory to his Creator? It simply means: I shall be beautiful before Him in observing the commandments. I shall prepare before Him a beautiful lulav, a beautiful *sukkah*, beautiful fringes and beautiful phylacteries.[17]

"Can you build a *sukkah* from an elephant?" (Drawing by Isaac Yallouz.)

Thus by furnishing the *sukkah* he portrayed so opulently, despite the rustic setting, Moritz Oppenheim was not just reflecting the values of a good, middle-class Jewish-German family—he was depicting this imperative to make the *sukkah* as beautiful as possible.

Commentaries   Nowhere in the description of the desert wanderings of Exodus are booths mentioned. Whereas the Torah does refer to clouds of glory. Commentaries often address such textual voids. For instance, leaving no assumptions unquestioned, the possibility that the booths were ephemeral, and not physical, is raised by Rashi. He comments on the verse: "'That I made the Israelite people live in booths'—this does not mean literally 'booths' but the 'clouds of glory' by which they were sheltered."[18]

Codes and Responsa   It turns out that in the context of the booths, the Talmud has pretty much laid the basics of the religious law. Still, later compilers of codes always had something to add. Here is Maimonides:

> Anything whatsoever is kosher to serve as the wall of a booth, for all that is required is a partition of some kind, even if it consists of a living creature. One may use a fellow human being for a booth wall on the first day of the festival. . . .[19]

In general, a human being, like an elephant, can serve as a wall. During the intermediary days of Sukkot there is no problem, but on the first day(s)[20] of the festival (governed by rules similar to those for the Sabbath) the human serving as a wall would be violating the prohibition against "forming a tent." However, if he is unaware that he is acting as a wall for your *sukkah*, then Maimonides permits this. Just don't let him get away!

Meanwhile, rabbis in hundreds of locations from Iran to Iberia answered questions posed to them by Jews. Rabbi Oshry's agonized responsum (cited earlier) is one example. These responsa were collected and today are more accessible than ever thanks to the existence of large-scale electronic compendia, such as the Bar Ilan University Responsa Project (which enabled us to find some of the responsa we cite in ensuing chapters).[21] On a computer with CD ROM in the Cornell library in Ithaca, New York, we can type in the keywords *sukkah* and *trailer* to communicate across centuries or continents to find out what responsa have been written on the question of a traveling *sukkah*. And most recently, e-mail has been used to consult distant scholars on pressing points of *halakhah*.[22]

**A time line of some of the people and events in Jewish history
mentioned in this book**

Years*	Events, Persons†
2000–1550 B.C.E.	Era of forefathers (Abraham, Isaac, Jacob)
1500–1000 B.C.E.	Exodus ca. 1280; conquest of Canaan by Joshua ca. 1240
1000–500 B.C.E.	King David d. 965; King Solomon d. 928; fall of first Temple 586
500–0 B.C.E.	Return from exile under Ezra ca. 450; Judah Maccabee d. 160
0–200 C.E.	Shamai/Hillel controversies; fall of second Temple 70; fall of Masada 73; controversy between R. Yehoshua & Rabban Gamaliel; R. Akiva d. 135
200–400 C.E.	Mishnah redacted ca. 210; Jerusalem Talmud redacted ca. 390
400–600 C.E.	Midrash Rabbah; Babylonian Talmud (Gemara) redacted ca. 500
600–800 C.E.	Period of Geonim in Babylonia
800–1000 C.E.	Midrash Tanhuma; R. Amram compiled order of prayers 860; Saadiah Gaon d. 942
1000–1200 C.E.	R. Gershom d. 1028; Solomon ibn Gabirol d. 1056; Rashi d. 1105; period of Tosafists; Abraham ibn Ezra d. 1164; Judah Halevi d. 1141
1200–1400 C.E.	Maimonides d. 1204; Nahmanides d. 1270; Zohar composed ca. 1286; R. Shlomo ben Aderet d. 1310
1400–1600 C.E.	Expulsion of Jews from Spain 1492; R. Yitzhak Luria (the Ari) d. 1572; R. Joseph Caro d. 1575
1600–1800 C.E.	First Jews settle in New York 1654; B. Spinoza d. 1677; Baal Shem Tov (founder of *Hasidism*) d. 1760; M. Mendelssohn d. 1786; R. Yehiel Mikhal of Zloczow d. ca. 1786; R. Elijah Gaon of Vilna d. 1797
1800–2000 C.E.	R. Hayyot d. 1855; R. Leiner publishes *tekhelet* treatise 1887; Moritz Oppenheim d. 1882; R. Samson Raphael Hirsch d. 1888; Theodor Herzl convenes First Zionist Congress 1897; David Wolffsohn d. 1914; H. N. Bialik d. 1934; R. Isaac Herzog d. 1959; R. Moshe Feinstein d. 1986; R. Joseph B. Soloveitchik d. 1993

* B.C.E. = Before the Common Era; C.E. = after the Common Era;
† d. = died; R. = Rabbi, Rav, or Rebbe.

Ethics, Aesthetics, Geometry, Poetry

Large sections of the oral law may indeed be legalistic. Not that the law is dry—one only need sit in the study hall of a *yeshivah* to see that. There is a great din—at tables young men sit facing each other surrounded by tractates of Talmud and commentary. They take turns, chanting a line of Mishnah, adding where its justification comes from, raising their voices for emphasis so that it would seem to a casual observer that they would soon come to blows, banging their hands on the table when done with their argument. Only at the end of their session, when they peacefully reshelve the worn volumes and go out talking about the chemistry lab do we see that the heated tone of their argument is just a piece of tradition. This is the way the oral law is studied.

The oral law is concerned not just with ritual obligations but with the totality of human activity, including ethical and aesthetic dimensions. For example, these considerations are embodied in the laws of the palm branch (*lulav*) whose waving is required during the Sukkot holiday:

> Mishnah: A palm-branch which has been acquired by theft, or which is withered, is not valid.

> *Gemara:* A withered one is invalid because it has no beauty; and a robbed one is invalid, because it is a *mitzvah* accomplished via a sin. . . .[23]

That the rabbis were concerned with the science and mathematics of their times is evident from the references to astronomy, botany, geometry, chemicals, and medicine. For instance, Talmud tractate Sukkah has a lengthy discussion about the area of a booth. Its minimum size is not much larger than a telephone booth. Then discussion moves to a cylindrical *sukkah*. Why? As we've seen, the reasons for seeming digression may be pragmatic, or they may be intellectual probing of the definition. The rabbis debate whether it is the perimeter or the area of the cross-section of the *sukkah* that should be conserved as the shape is modified from rectilinear.

> Consider: by how much is a square greater than its (inscribed) circle? By a quarter. . . . That is so in the case of a circle inscribed in a square . . . but if a square is to be inscribed within a circle a

greater circumference is required on account of the projection of the corners. But consider: if the side of a square is a cubit [about 22 in.], its diagonal is approximately one and two fifths cubits. Should not then [a circumference equivalent to] sixteen and four fifths cubits suffice?[24]

The page of Talmud reproduced below shows the use of little drawings to make the point; the reader should not get the idea that

A page of Talmud that discusses the dimensions of a *sukkah*, and illustrates the discussion with geometrical drawings.

such illustrations are common, for they are actually rare. But the feeling that is generated is interesting, and there are parallels to the imperative to show by drawing in a chemical paper.

The discussion in this passage is about circles and squares, not actual booths. Relevant are the formulas for the diagonal of a square (if the side of a square is 1, then its diagonal is $\sqrt{2} = 1.414\ldots$), and the value of π, the ratio between the circumference of a circle and its diameter. You will see in this passage that the sages approximated the irrational number $\sqrt{2}$ by 1.4, which is not bad. They did much worse with π, which they took as 3.

The reason we say "worse" is that Archimedes had derived a much better approximation several hundred years before, and this value was known in the Middle East. The Talmud specifically acknowledges in another place that in astronomy the calculations of the "sages of the nations" (i.e., non-Jews) should prevail.[25] Similarly, the medical opinions of the Talmud are not usually held as authoritative today.

Although the walls of the *sukkah* must have a certain geometry, they can be made of nearly anything. But the roof must be made of *s'khakh,* a word easier to pronounce in Hebrew than it seems in the transliteration (a possible translation is "thatch"—just as much fun to pronounce). *S'khakh*—cut vines, branches—must not be put on too thickly. In the midst of that dry legal account the Talmud brings in a twinkling of poetry:

> [It's preferable] the roof thatch be loose enough so that the large stars can be seen. . . .[26]

Sukkah and Demarcations of the Natural

Let us return to the Moritz Oppenheim painting. What is natural about this *sukkah* and what is unnatural?

The family pictured is hardly back-to-nature. From floor to ceiling the *sukkah* is festooned with the refinements of the bourgeois family it houses. No leaf or stray bit of nature touches the lush carpet that covers the wooden floor. The drapes, no flimsy affair, are the very same heavy satin curtains that appear in the other scenes Oppenheim painted of the family's regular dining room. The weighty chandelier would be equally in place in a ballroom. The mahogany-framed painting on the wall, the fine linen tablecloth, the hand-painted china, and the holiday finery remind us that this is no picnic in nature.

In fact, the whole idea of building a *sukkah* is unnatural. [Here we go again, as if our correspondence with Ayyal in Chapter 1 were not enough.] When we celebrate the holiday today, we erect a patently synthetic imitation of the booths that were said to have sheltered the Israelites during their desert wanderings. The only intrusion of nature is the *sukkah's* roof, which is made of boughs. Everything else is unnatural—that is, made by human hands.

In contrast to the idyllic scene Oppenheim portrays, the reality of eating in a *sukkah* is not always so attractive. A rabbi from Phoenix, Arizona, where autumn temperatures can surpass 100° F, reports that the question most frequently posed to him was: "Is it kosher to air-condition the *sukkah*?" At the other end of the thermometer, it has been told that a U.S. Air Force chaplain in Anchorage, Alaska, built *sukkah* walls from igloo-like ice-blocks!

What these two structures, from Arizona and Anchorage, have in common with each other, and with Oppenheim's *sukkah*, is the roof of branches. The etymological root of *sukkah* is *s'khakh*, meaning covering or protection, a thatched roof. Thus, for the essence of the *sukkah*, and of the entire holiday, we must look up to the roof. Here we gain insight into how tradition differentiated natural from unnatural, and perhaps find an answer to our question about Oppenheim's *sukkah:* Where does the natural leave off and the man-made begin?

Whereas there is carte blanche for walls, there are three strict requirements that dictate which materials may be used for the roof. Those were stated in the Mishnah quoted on page 64, beginning "If he trained vines." Underlying these dry laws we find a complex philosophical approach to the question raised above: "What is natural and what is unnatural?"

The thatch covering the *sukkah*, to be ritually valid:

1. must have grown from the ground;
2. must be cut off from its roots; and
3. must be incapable of becoming defiled (ritually impure).

The first two requirements imply that the roofing materials must originate in nature but may not be used in their natural state. These rules form the lower boundary of the spectrum of appropriate materials. A growing vine trained across a booth top is too natural. It is not ritually valid because human beings did not make it. We are enjoined in Deuteronomy 16:13: "You shall *make* the Sukkot festival. . . ." A growing vine is nature-made and is not a fulfillment of this injunction ordering us, humans, to "make" the festival. "Making" entails cutting and placing. Cutting a vine that had been trained over the *sukkah* is not enough. After being severed, each vine has to be lifted and set down again so that *we* have done the making, not nature.

But lest we go too far in rendering the thatch human-made, an upper boundary is stipulated. The third rule insures that the thatch will not become too human-made and lose its natural properties. It is phrased in the language of the talmudic laws relating to ritual purity and impurity, but the implications are startlingly modern. To understand the third thatch requirement, we must take a short excursion into the laws of ritual purity, because they hint at a demarcation between natural and unnatural.

Rabbi Adin Steinsaltz points out that "the laws of purity are essentially a complex, unified network of laws, interrelated within a special logical structure."[27] The Torah offers no explanations for purity laws and it is risky for us to do so. Furthermore, since the destruction of the Temple in Jerusalem many of the laws are inapplicable because they had been related to the Temple ritual.

Ritual purity and impurity are not concepts of physical cleanliness or hygiene, as can be seen from the fact that ritual hand washing must be preceded by a regular washing.[28] In general, what is living and healthy contains no impurity, and impurity increases with proximity to the ultimate lack of freedom: death. Impure sources—a corpse, carcasses, humans in certain conditions—can transmit their ritual impurity to objects that come into contact with them. Many things cannot become impure—for example, bodies of water, living animals, growing plants, and unfinished objects. To categorize any given object, one must take into account its source material, shape, and intended use.

Source material: Utensils made from unbaked earth, stone, or marble are impervious to ritual defilement. This is one reason archaeologists find so many stone objects in Jerusalem excavations.

On the other hand, objects made from wood, metal, leather, bone, cloth, sackcloth, and baked clay can become impure. Glass is classified with metal because both materials are recyclable. And our "unnatural" plastics? These need careful consideration; much recent responsa literature addresses use of the newer materials.[29]

A pattern seems to emerge: Materials that are further from their natural state and undergo more transformations are more susceptible to impurity—for example, metal has to be mined from ore, smelted, and then shaped, whereas stone can be hewn directly.

Shape: Concave wooden objects can hold liquids which are susceptible to impurity. Nonconcave wood, which is closer to its natural state, for example, branches, palm fronds, and slats, are valid for roofing since they are impervious to impurity.

Intended use: If you come across a wood or reed mat that seems to fit the bill for thatch there may be a problem. Mats made for reclining are ruled by the Talmud (after substantial discussion) to be susceptible to impurity, whereas mats intended to provide shade are impervious (and valid for *sukkah* roofing). We can't submit all mat makers to a polygraph test to gauge their intentions. So the codes provide guidelines based on assumptions about the majority of people in a given city: Small mats are usually for reclining and therefore not fit, but large mats, usually for shade, may be used. (Bamboo mats of the latter type, popularly sold for roofing in Israel, are called "thatch forever" which is an oxymoron since by definition a *sukkah* is a temporary dwelling!)

What emerges from the laws of Sukkot is that the talmudic minds some two thousand years ago were deeply engaged in dividing the world around them into categories. At first these might seem bizarre, but their schemes of categorization shed light on our present-day concerns about the amount of artifice and the synthetic in our lives. In recent decades there has been renewed interest in issues concerned with nature versus technology, and a *Zeitgeist* including Green politics, deep ecology, New Age science, and ecofeminism has emerged. In an attempt to address these concerns, a group of young Jewish activists formulated a code in the 1960s for readers of the *Whole Earth Catalogue*. Called *The Jewish Catalogue: A Do-It-Yourself Kit*, this code popularized Sukkot as an *in* holiday:

> *Sukkah*-building: if you can get into this mitzvah you will probably find great joy in it. . . . Place some 1 × 1's running in both directions on the roof and cover that with rushes or pine boughs. The entire roof must be made of organic material.[30]

Note the buzzword, "organic," which reflects the authors' concern about our over-reliance on the "unnatural." By condensing all the laws about the roof cover into that one word, brevity has been gained. But their definition is intellectually poor in comparison to the sensitive and profound discussion of Sukkot in the Talmud, which reveals that these questions about natural and unnatural were addressed by the framers of Jewish law millennia ago.

Why We Like the Natural

The difficulty in disentangling the natural from the unnatural is not a problem only for religion. Scientists, especially chemists, often feel beleaguered by society because they produce "unnatural" and often downright dangerous materials. Whereas the terms "natural," "organically grown," and "unadulterated" have positive connotations, synthetics seem at best conditionally good. Chemists are quick to argue that "natural" objects cannot really be distinguished from "unnatural" ones, for example, Vitamin C extracted from natural rosehips is identical on a molecular level to Vitamin C produced synthetically in a lab (see Plate 7). Nevertheless, the distinction between natural and unnatural has a hold on our psyches in daily life. Why is it that we often seek out the natural, whether we are discussants in a talmudic debate, or readers of *The Jewish Catalogue*, or chemists manufacturing synthetics?

There are psychological and emotional forces at work in determining our preference for the natural. Some of these factors can explain the attraction of the natural even for scientists who know the value of the artifactual and who perceive the lack of deep difference (at the molecular level) between the natural and the synthetic.

One factor is romance, an unrealistic striving for what no longer is or cannot be. This probably accounts for the popularity of the nostalgic painting of a *sukkah* by Oppenheim. There is a certain irony in the fact that this very painting, indeed every *sukkah*, is an unreal, unnatural, but entrancing attempt to replicate the natural. Such romantic paintings have a hold on us that is stronger than reality because of the image in our minds. A reaching out for nature, for real wood, the smell of hay, the feel of the wind in the sails, still determines our desires. Our penchant for romance extends to other areas. It doesn't matter that old train stations were nasty, filthy buildings. When we think of an old train station, we see Ingrid Bergman saying good-bye to Leslie Howard, and that scene forms an image in our minds of what a train station should be like.

Similarly, it doesn't matter that feeding a large family cramped in a dingy *sukkah* on a damp, windy night is no picnic; our mind's *sukkah* is just right.

Another reason we are attracted to the natural is the alienation we feel when our circuits are overloaded with the unnatural and synthetic in the environment. Sometimes the superabundance of artificial objects repels us. The typical American motel room, for instance, offers little respite from the artificial. The variety of plastics and synthetic fibers in the furnishings is astonishing and even intellectually interesting, as a basis for a course in polymer chemistry. But one is hardly attracted to the setting. We are distanced from our tools, and from the effects of our actions. We see it in routine work on an assembly line, in selling lingerie, even in scientific research. We work repetitiously on a piece of something, not the whole. But there is something deep within us that makes us want to see the signature of a human hand on a product; sometimes the *Jewish Catalogue* does get it right:

> One of the good things about a *sukkah* is that you should build your own. Even if you buy the prefab variety, you should erect it yourself. Most of us live in houses or apartments built by others. Most of us eat bread baked by professionals. Like *hallah*-baking, *sukkah*-building gives us the chance to enjoy the fruits of our own labor.[31]

A third factor that makes us seek out the natural is spirit, an innate need for the surprise, the unique, the growing that is life. A pine struggling to grow in a Jerusalem cleft can send our thoughts forward in time to when it, or its offspring, will eventually split that rock. Or it can send our memories decades backward to its initial planting. This associative capacity of the soul is at work when we sit in the booth. The *sukkah* is supposed to trigger in the collective historical memories of Jews a recollection of God's grace in providing booths for the Israelites after they left Egypt: "in order that future generations may know that I made the Israelite people live in booths. . . ."

The flimsiness of the *sukkah* is designed to remind us of our own vulnerability to nature, which is neutral but often inimical to anthropocentric us. This is embodied in a law we did not mention earlier, that rain must be able to penetrate the thatched roof.[32] It is hard to appreciate God's protection while ensconced inside our concrete-and-steel homes.

A booth may be a jerry-built hut that looks like the work of a well-intentioned ten-year-old. Or it may look like Moritz Oppenheim's painting, something straight out of a nineteenth century *Better*

Homes and Gardens. But to be a valid *sukkah*, it must meet the requirements outlined in the law codes. (See Plate 8.)

A Building Manual?

The philosophical underpinnings of the holiday will be found, *inter alia*, in the petty details of booth building. That the Talmud and later codes are not construction manuals is seen in the story of the *yeshivah* student who set out to build his first *sukkah* for himself and his bride. Not knowing how to begin, he asked his teacher.

The rebbe said: "Learn the first two chapters of Tractate Sukkah, then go build!"

The first night of the holiday, the *sukkah* came tumbling down. Crestfallen, the groom ran to ask his rebbe why his booth had collapsed. The rebbe opened his Talmud and said:

"You ask me, 'Why did the *sukkah* collapse?' Too bad you didn't read Rashi's commentary on the next page *before* you built your *sukkah*. He asks the same question!"

No, what we have in the traditional texts is not a builder's handbook,[33] but a sensitive, multilevel inquiry into issues, such as natural and unnatural, that refuse to go away.

You Must Not Deviate
to the Right or the Left

I

What moved God to labor as He did for the six days of Creation? Perhaps it is not for us to know, for the question of motive is a very human one. But we perceive a hint in the marvelous story of Creation, in the insistent theme of *separation*—of light from darkness, of water below from water above, of day from night, of woman from man. Difference, the necessary consequence of separation, is at the root of interest, and ultimately of beauty. There is difference galore in the subsequent happenings of the Bible.[1]

Difference, or better said, the tense edge between similarity and difference, is also at the heart of chemistry.[2] Salt, sugar, penicillin, and sodium cyanide are white crystalline substances; their differences emerge in taste, toxicity, the conduction of electricity by their solutions. The atoms making up these molecules are not the same, no matter how similar the substances appear to the "naked eye." Our instruments measure not one, but many telltale differences. And the body certainly knows.

"Subtle is the Lord," said Albert Einstein,[3] a nicely unreligious scientist who delighted in using religious discourse. It turns out that you don't even need a different set of atoms to manifest critical variance. Molecules exist that are composed of identical atoms, the atoms connected to each other in the same (well, not quite the same) way. Yet they are different, these molecules, because they are "handed," and while most of their properties are identical, they may differ in subtle, yet biologically significant ways.

What does it mean that a molecule is "handed"? That it has the characteristics of a left or right hand. But what characteristics? Does it *look* like a hand? No, that's just a metaphor we use; the geometrical essence of a right or left hand that a molecule possesses is what matters.

Let us look at the essence of handedness. Right and left hands are certainly similar but not identical. Neglect fingerprint differences, and those lines in our palms that some say bear fortune, and warts, freckles, and the scar that gives evidence of a slipped knife, the asymmetrically bitten nail. Take the Platonic ideal of a left hand, and it still differs from the right in a way that we have no trouble recognizing in others (shake hands with the two options and you'll be quick to see it!) but that we may have difficulty in verbalizing for ourselves. Until . . . we look in a mirror, and perceive that the seemingly life-like similacrum is brushing his or her teeth with a left hand, while we (we check this a little nervously) do so with our right.

A left hand—the idealized left hand—is a mirror image of a right hand. It is neither identical nor superimposable on a right hand. By superimposable we mean you cannot put your left hand on your right (or vice versa) while both palms are downward and have them match; they are subtly different in just that mirroring way.

Some objects *are* superimposable on their mirror images. The gold band on your finger, the baseball cap (without the writing on it), most of the things we wear (not our shoes, and ignoring the fasteners on shirts, blouses, and pants) would fit the mirror-image creature just as they fit us. But some things—the body, shoes, and shirts if we worry about the fasteners—exist as nonsuperimposable mirror-image pairs. Obviously a certain spatial complexity is required to achieve the potential differentiation of an object and its mirror image. What is that complexity?

It is easier to explain the difference for a molecule than for a hand. And still easier to explain with ball-and-stick chemical models. But the persevering reader lacking three-dimensional aids can understand what follows; we think it could even be read in bed without a pencil. Taking a cue from another chemist, Harry B. Gray,[4] we had a colleague test this out by retiring with a draft of our book on two successive evenings. He was able to follow the discussion in this chapter even though his wife would not allow him to take any models to bed to aid his study.

Take the prime building block of organic chemistry (and therefore of most biological molecules), the carbon atom. It bonds effec-

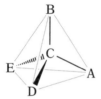

The four bonds that carbon normally forms
point toward the corners of a tetrahedron.

tively to as many as four other atoms. These may be hydrogen; halogens such as F, Cl, Br, or I; oxygen- and nitrogen-containing groupings; and other carbons. The bonds to these atoms are not randomly disported in space, but go off in "tetrahedral" directions from the central carbon—as from the center of a tetrahedron to its vertices (see the illustration above). This is carbon's geometrical essence; if the atoms bonded to carbon are all the simplest, H, then this is the archetype of organic molecules, methane, CH_4, the main constituent of natural gas. In the figure we use a chemist's notation, emerging from a desperate struggle by a group of artistically untalented people to communicate essential three-dimensional information: A solid line is taken to lie in the plane of the paper, a wedged line comes to the front, and a striped wedged line goes behind the paper. C is the central carbon atom; A, B, D, and E, the "substituents."

If all four, three, or even just two of the four atoms or groups of atoms bonded to a carbon are identical, the molecule that results (CA_4, CA_3B, CA_2BD—all of these "substituted methanes") is indistinguishable from its mirror image. Try it!

However, if the four substituents around that central carbon all differ (see the illustration at the top of page 82), then no rotation (please, please try!) will bring all atoms into congruence: The two CABDE molecules shown are nonsuperimposable mirror images. They are called enantiomers. Molecules capable of showing such handedness are said to be chiral (from the Greek *cheir*, for hand). Chirality is exhibited not only by appropriately decorated tetrahedral structures; screws and helices also are inherently chiral. Textile specialists could puzzle out the origins of the wool and linen cloths found at the ancient Israeli fortress of Masada on the basis of the

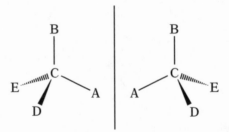

The nonsuperimposable mirror-image forms of a CABDE molecule.

twist of the spun threads—whether they are "S spin" or "Z spin" (see the illustration below).[5]

But in what way is a human hand like a helix or a tetrahedron? A hand just *looks* different, so much richer in detail, than a carbon atom. Imagine yourself sitting at a point near the center of your lifeline, maybe inside the flesh. Then think "palm, back, pinkie, thumb," that is, think where these parts of the hand are in space relative to you (top, bottom, left, right). You have described a set of four noncoplanar indicators (see the left side of the illustration on page 83[6])

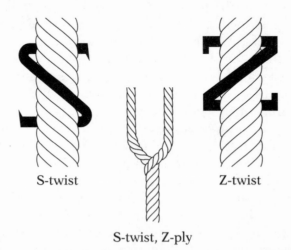

S-twist

Z-twist

S-twist, Z-ply

Textile scientists define the two chiral ways (left and right) of spinning a thread as "S" and "Z." Similar notations apply to the "ply," the way the threads are twisted to make a stronger yarn.

(*Left*): a schematic correspondence between the essential descriptors of any hand and a tetrahedron; (*Right*): A well-known hand.

that are formally like the labels of the four substituents on a chiral carbon atom.

Why do we care about the difference between left- and right-handed molecules? Our left and right hands are very similar—they weigh the same (think of mannequins, not people), they have much the same color. But now put your hands to work. Imagine the difference between using your left or right hand to shake the hand of a person proffering a right hand. Or imagine turning a left- or right-hand threaded screw to the left into a nut of a certain matching handedness.

So it is with molecules, because they don't sit still, they do things to us. Morphine, to which all of us who have ever had an operation are grateful (and which is strongly addictive), is a chiral molecule. One of its mirror-image forms makes us love and fear the molecule; the other enantiomer is a much less effective painkiller and is also less addictive. Aspartame is a modified amino acid that is 200 times sweeter than sugar. Its enantiomer tastes bitter. There is a patent for another sweetener made from a left-handed sugar which is sweet but indigestible (because our bodies only digest well right-handed sugars), therefore nonfattening. One form of limonene smells like oranges, its enantiomer like turpentine. A bark beetle, *Ips pini*, is attracted to the eponymic chemical ipsdienol; the mirror image form of the molecule has no effect on the bug.[7]

An American poet asks the next question:

> . . . *you'll say: What's left and right*
> *got to do with it, do molecules*

*hide switch-hitters, or the lovingly
taught fit of small arms and sleeves?[8]*

Why the difference in the properties of left- and right-handed molecules? It takes a hand to know one—the molecules in our bodies, simple and complex as they are, the molecules that recognize what we eat or what is injected into us, are often handed. In fact, all the proteins, those wonderful chemical factories that disassemble food into molecular pieces, only to reassemble the fragments elsewhere in the body into hair, skin, heart tissue, and sperm, the proteins that also do all the work of transport and defense, are built up of amino acids of one specific handedness—they're all lefties.[9] (See the illustration below.[10]) They curl up, these proteins, into globular equivalents of a glove, a messy-looking tailoring job of a glove that is most certainly left-handed and that either fits very well, or doesn't fit one enantiomer of morphine or limonene or aspartame.[11]

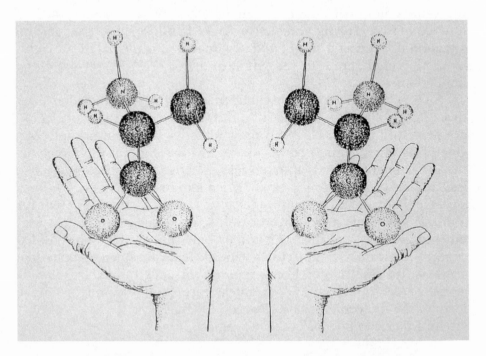

Right- and left-handed molecules of an amino acid, alanine. (Drawing by Ronald N. Bracewell.)

Plate 1
"Tnuva Dairies Bring You the Finest, Straight from the World's Best Production Line. Nature!"

תנובה מביאה אליכם את כל הטוב, היישר מפס הייצור הגדול ביותר בעולם. הטבע.

Plate 2
Moses and the Burning Bush. The sheep are in the background in this fourteenth-century Spanish Haggadah illustration. John Rylands University Library, Manchester.

Plate 3
The Young Bull *(1647) by Paulus Potter reflects
a shift in focus to nature in foreground,
dominating the scene. Mauritshuis, The Hague.*

Plate 5
*A page from the Yaffa Wigs
catalogue. Human hair wigs can
be twice as expensive as those
made from synthetics. There are
also blends.*

Plate 4
Travelers on a
Mountain Path, *by
Fan K'uan, early
eleventh century.
Chinese artists
focused on
landscape
centuries before
Europeans did.
National Palace
Museum, Taipei.*

Plate 6
Conundrum. One of the authors perusing the offerings in the window of the Yaffa Wigs shop in Israel. Note the intrusion of the natural toward the top of the photograph.

Plate 7
An advertisement for Solgar vitamins features a molecule artfully contorted in the shape of a luscious bunch of grapes. Not many ads today feature a molecule, so perhaps we should be grateful (or grapeful). But this molecule, of course, is being set up to be the villain. The ad headline says: "Maybe you don't know that you are swallowing synthetic vitamins, but your body knows" and concludes, "Solgar, the world's most natural vitamins."

Plate 8
A student in Kiryat Sanz, Netanya, leaves the study hall to help build the Rebbe's communal sukkah, which accommodates hundreds.

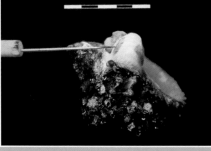

Plate 9
The three muricid species used in the ancient dye industries; below them a needle points to the hypobranchial gland of a Thais haemastoma specimen. The scale is in centimeters.

Plate 10
Old tekhelet, new flask: Three young people (members of a group reviving the tekhelet industry in Israel) instructing a theoretical chemist in the art and science of dyeing wool with tekhelet from Murex snails. The process is being carried out in a Phoenician dyeing pit at Tel Dor. One of the instructors said, "Who would have imagined that the chemistry we so reluctantly took at Yeshiva University would be of any use?" Compare the blue of the wool, emerging from the yellow liquid, with the sea and the sky.

II

Concern about left versus right is hardly limited to science. In Deuteronomy 17:8–11, in a section on judges, the following passage appears:

> If a case is too baffling for you to decide, be it a controversy over homicide, civil law, or assault—matters of dispute in your courts— you shall promptly repair to the place which the Lord your God will have chosen, and appear before the levitical priests, or the magistrate in charge at the time, and present your problem. When they have announced to you their verdict in the case . . . you shall act in accordance with the instructions given to you and the ruling handed down to you; you must not deviate from the verdict that they announce to you either to the right or to the left.

The intent of this passage seems clear—the decisions of the priests and judges *must* be obeyed. The people are admonished to do so, in no uncertain terms.

What caught the attention of one of the authors, Shira Leibowitz Schmidt, was the formulaic terminology at the end of this passage: ". . . you must not deviate from the verdict that they announce to you either to the right or to the left."*

Shira said: "Hmm. I wonder if they had chirality in mind."

To which Roald replied: "You're crazy, Shira; they're just using a figure of speech, telling people that they had better adhere to the word of the court, period."

"Well, I'm not so sure," Shira said, reaching for her commentaries. "First of all take Rashi. On this passage he writes:

> 'To the right or to the left' even if the judge tells you about what appears to you to be right that it is left, or about what appears to you to be left that it is right, you have to obey him; how much the more is this so if actually the judge tells you about what is evidently right that it is right and about what is left that it is left.[12]

---

* What follows is the first of several "episodes." These will highlight cases of authority (asserted, disputed) in science and religion or elsewhere, as in this first one. . . .

"It seems to me that Rashi is worrying about how to tell right from right and right from left. Isn't that what chemists do?"

"I hope the syntax of that Rashi commentary is clearer in Hebrew. But I still say you're way out, Shira—why bring in stereochemistry (several millennia in the future), when all these guys are doing is using a metaphor?"

"Sure, Rashi is expanding the biblical metaphor. But it's in the nature of talmudic and post-talmudic debate that everything is real, and everything is at the same time metaphorical . . . and sometimes anachronistic."

Shira went on: "Or to put it another way, the discussion may be about obeying the courts, but it is also about chirality. Anyway, didn't you once tell me about Pasteur having to convince Biot that he, Pasteur, could separate enantiomers? And what about that tirade of Kolbe on that upstart van't Hoff? It seems to me that the story of chirality is full of people telling each other that left is right!"

III

The story of Pasteur *is* worth telling.[13] When chirality was first broached, no one in the world had any inkling of the tetrahedral carbon atom. At the beginning, which was early in the nineteenth century, one had only curious observations on solids and solutions rotating the plane of polarized light.

Here comes some physics: Normal light, unpolarized, is made up of oscillating electric and magnetic fields. With the aid of reflection from certain surfaces, or later with special filters (which we find commonly today in depolarizing sunglasses as well as certain airplane windows), it's possible to isolate a certain special kind of light. It has all the colors of normal light, but in it the electric and magnetic fields each oscillate in one and only one plane. This is called polarized light. You can think of the polarizing filter like a slit, letting through a tape-like form of the light.

Polarized light was much studied in France. François Arago, an astronomer, found in 1811 that a quartz crystal cut in a special way took "the plane of polarization" of light and rotated it by a certain angle. This could have been something peculiar to the solid, but then Jean-Baptiste Biot, in 1825, found that solutions of some naturally occurring substances also exhibited this property of "optical rotation." This was significant, because the property, whatever it was, had to reside in the tiniest chunks of matter dissolved in the liquid, which people were only then recognizing as molecules. Sometimes the solutions rotated the plane of polarized light to the right (they were dextrorotatory), sometimes to the left (levorotatory). The illustration below shows schematically how some solutions take the polarized light and rotate the plane of polarization. Pretty incredible, even today.

Enter the young Louis Pasteur, rabies vaccine and milk sterilization a quarter of a century in the future. In 1848, at the age of 26, he studies a curious chemical compound synthesized in wine-making, racemic acid. Its chemical formula is (and was then known to be) identical to that of another more common acid, tartaric acid, which is deposited (as the potassium salt) on the walls of wine barrels, and occasionally in white wine bottles after fermentation. Tartaric acid is "optically active," that is, rotates the plane of polarized light. Racemic acid, seemingly identical to it in composition, is not. What is going on?

Pasteur, only 26, examines salts of racemic acid. His powers of observation are keen. He crystallizes one such salt, the sodium ammonium one, and sees lovely crystals (each we now know containing billions of billions of molecules, each of the same handedness). The crystals have finely developed facets and they differ in a

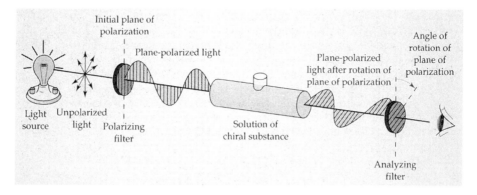

A schematic diagram showing first how a "polarizing filter" takes ordinary light and allows only light whose electric field is polarized in a certain plane to come through. The solution of a chiral molecule then rotates that plane of polarization, either to the left or to the right. A second polarizing filter is used to determine by just how much that plane of polarization was rotated.

very subtle way, as a left hand from a right. The crystals are as mirror images; their shapes are shown below. Pasteur separates them, and finds that a solution of one crystal rotates the plane of polarized light in a direction opposite to a solution of the other crystal. One is identical to tartaric acid, the other is its hitherto unknown enantiomer. The one-to-one mixture of the two was racemic acid.

Mirror-image crystals of sodium ammonium tartrate, synthesized by Pasteur himself, and a drawing of the mirror-image crystals.

Pasteur was lucky. Most substances that are an intimate mixture of enantiomers do not spontaneously separate into crystals that contain only left-handed molecules, and others that only contain right-handed ones. Just a few instances of this happening (other than racemic acid) have been found since Pasteur's time. And the molecule he was studying does that separation only in a narrow temperature range. He was very, very lucky.[14]

He then had to convince others that what he had done was right. He especially had to convince the doyen of French investigators of optical rotation, the aforementioned Biot. Here is what Pasteur writes about his experience:

> The announcement of the above facts naturally placed me in communication with Biot, who was not without doubts regarding their accuracy. Being charged with giving an account of them to the Academy, he made me come to him and repeat before his eyes the decisive experiment. He handed over to me some paratartaric acid [another name for racemic acid] which he had himself previously studied with particular care, and which he had found to be perfectly indifferent to polarized light. I prepared the double salt in his presence, with soda and ammonia which he had likewise desired to provide. The liquid was set aside for evaporation in one of his rooms. When it had furnished about 30 to 40 grams of crystals, he asked me to call at the Collège de France in order to collect them and isolate before his eyes, the right and left crystals, requesting me to state once more whether I really affirmed that the crystals which I should place at his right would deviate [the plane of polarized light] to the right, and the others to the left. This done, he told me he would undertake the rest. He prepared the solutions with carefully measured quantities, and when ready to examine them in the polarizing apparatus, he once more invited me to come into his room. He first placed in the apparatus the more interesting solution, that which ought to deviate to the left. Without even making a measurement, he saw by the appearance of the tints of the two images, ordinary and extraordinary, in the analyzer, that there was a strong deviation to the left. Then, very visibly affected, the illustrious old man took me by the arm and said:

"My dear child, I have loved science so much throughout my life that this makes my heart throb."[15]

Biot is moved, the authority convinced by the miracle that
repeats itself so many times—the beautiful and unexpected but
reproducible experiment. The act of creation whose beauty resides
in the fact that anyone—no, everyone—can do it.[16]

IV

Rashi, the great Franco-Jewish exegete whom Shira cited, owned
vineyards, so he is likely to have seen crystals of tartrate.[17] But back
to religion: Rashi may have simplified things a little in that passage
quoted above, about obeying, "even if the judge tells you about what
appears to you to be right that it is left." For the Jerusalem Talmud
takes quite a different tack from Rashi in a comment on the very
same passage in Deuteronomy:

> You might think that if the Sages tell you that the right is left or the
> left is right that you are to heed them. The text states: "to the right
> and to the left"—when they tell you the right is right and the left is
> left.[18]

So . . . do not accept the court's judgment if it is wrong.
Nahmanides tries to find a compromise. He accepts the possibility that the courts might err (indeed, nearly a whole tractate of the
Talmud is devoted to procedures when a court's judgment is erroneous). But Nahmanides makes an exception for the great Sanhedrin,
the highest court of them all. The High Court's judgment must be
accepted if the social contract and the compact between God and
man is to stand:

> The need for this commandment [to follow the rulings of the
> Sanhedrin] is very great since the Torah was given in writing.
> Now it is well known that no two people can agree on all
> matters that arise. Controversy is thus bound to increase and
> the Torah will be subject to various interpretations. Scripture
> therefore laid down the law that we should obey the Great
> Sanhedrin that sits before the Lord in the place which He shall
> choose, whatever they instruct you in their interpretation of the
> Torah, whether it was transmitted to them in unbroken succes-
> sion by oral testimony, or whether it is based on their under-

standing of the sacred text. For the Torah was given on condition they were vested with its interpretation, even if they seem to call the right left. How much more so should you give them the benefit of the doubt and accept their right as right; for the spirit of God rests on the ministers to His sanctuary. He will never forsake His kindness and always preserve them from error and stumbling.[19]

In the passage just preceding this comment on our verse, Nahmanides cites the wonderful story of Rabbi Yehoshua and Rabban Gamaliel, which it behooves us to tell as well.[20] Rabban Gamaliel, the head of the Sanhedrin, was at loggerheads with his teacher, Rabbi Yehoshua, the preeminent scholar of late first century Palestine. There is an important calendric computation of Yom Kippur, the Day of Atonement, based on a critical observation of the moon rising. The two rabbis made their calculations, and Rabbi Yehoshua, senior in scholarship, said that the Holy Day would fall a day later than the computation of Rabban Gamaliel indicated. Mishnah Rosh Hashana continues:

> Thereupon Rabban Gamaliel sent to him [Rabbi Yehoshua] saying: "I command you to appear before me with your staff and your money [forbidden on the holiday] on the day which according to your reckoning should be the Day of Atonement." . . . Rabbi Yehoshua then went to R. Dosa who said: "If we call in question the decisions of Rabban Gamaliel, we must call into question the decisions of every Court which has existed since the days of Moses to the present." . . . R. Yehoshua thereupon took his staff and his money [with great personal anguish] and went to Yavneh to Rabban Gamaliel on the day which he [Rabbi Yehoshua] calculated was Yom Kippur. Rabban Gamaliel rose and kissed him on his head and said to him: "Come in peace my teacher and my disciple—my teacher in wisdom and my disciple because you have accepted my decision."[21, 22]

There is substantial agony in this personal tale, as there is, we think, in the writings of most rabbinical authorities on the application of the right-left passage. It is difficult to deal with error, yet, to quote a later sage, "to err is human. . . ." Even for the seventy-one sages of the Sanhedrin, as for the nine of the Supreme Court of the United States.

As the Jews were forced into distant exile, making communication with other Jewish groups difficult, and as these communities faced different material and spiritual exigencies, there grew the acceptance of rabbinically sanctioned local variations in practice. The Sanhedrin no longer sat. One respected the religious thinking of scholars (of which we will have more to say), especially of those who earned by popular acclamation the designation of being called "the great of their generation." But even the greatest of religious scholars recognized that their authority extended at best to only their own community.

An interesting case in point is the history of rulings on polygyny. The practice is explicitly sanctioned in the Bible, as we see for the patriarchs. Actually polygyny was rare in post-biblical Jewish communities. In the Middle Ages an outstanding scholar of Mainz, Rabbenu Gershom (d. 1062), issued a ban on polygyny. This was accepted as binding on European Jews, but explicitly invalid for Sephardic Jews, some of whom continued to practice polygyny (as their surrounding Muslim neighbors did) until recent times.[23]

What about rabbinical authority today? Within Israel, there has arisen much tension because various diaspora communities, whose practice had evolved in distinct ways, now were forced into close geographical proximity in the newly reconstituted state. If we follow the aforementioned Sephardic and European Ashkenazic communities to Israel we can see how some differences were worked out. Would each group maintain, for example, its own marriage customs? Or would one adopt the custom of the other? In the case of the issue of polygyny, Rabbenu Gershom's ban prevailed over all communities (although the minuscule number of immigrants who came from countries where polygyny was allowed maintained their family framework).

There were disparate traditions not only between the Sephardic and Ashkenazic groups, but a sizeable nonobservant proportion (mostly European immigrants) rejected the authority of the rabbis and the texts. The result is a major schism in the country between observant and nonobservant Jews, in addition to some differences within the religious fold.

At the same time we see a reaffirmation in practice of the community's acceptance of diversity. At least sometimes. During the Passover holiday on an Israeli army base (where there may be even some non-Jewish, Bedouin and Druze soldiers), everyone good-naturedly eats the *matzah* that replaces bread for a week (but not one hour more than necessary!). The food officially provided on the base follows the more lenient Sephardic custom, whereby all wheat products are banned but legumes and rice are allowed. Ashkenazic soldiers can abstain from the legumes and rice, thus following their custom. And those who choose not to keep Passover food strictures at all can be found foraging . . . but off base.

There are many issues much more acrimonious and divisive than these. These groups vie politically, not only for representation, but also for setting the state agenda on important religious issues: Who is a Jew? Shall buses run on the Sabbath? Shall Orthodox fulltime scholars receive deferments from military service? A fragmented political system in a secular state that avowed spiritual continuity to ancient Israel provided an opening for the relatively small observant fraction of Israeli Jewry to exercise an unusual political role.

V

How *does* one distinguish between left- and right-handed molecules? Pasteur's first way—of separating crystals of the twain—works very, very rarely. He himself developed two further techniques. First he found that a mold, *Penicillium glaucum* (related to the molds that many years later gave us penicillin), grew on a nutrient solution that contained optically inactive racemic acid (by then known to be a mixture of the two mirror-image molecules). Remarkably, the solution left behind after a period of mold growth rotated the plane of polarized light to the left. The mold metabolized the dextrorotatory isomer preferentially.[24]

This method of separation works sometimes, but depends on a cooperative bug, and it does away with half the stuff. One can do better. Still making use of naturally occurring handed molecules, chemical "bases" this time, Pasteur finds a general reaction of salt formation of such a base, call it L-base, with a mixture of the enantiomers of some acid. Suppose the mirror-image acid molecules are called D-acid (D for dextrorotatory) and L-acid (L for levorotatory). The reactions may be written as follows:

$$D\text{-acid} \quad + \quad L\text{-base} \quad \rightarrow \quad D\text{-L salt}$$
$$L\text{-acid} \quad + \quad L\text{-base} \quad \rightarrow \quad L\text{-L salt}$$

Now the D-L salt and the L-L salt are *not* mirror images. They are to each other as the "union" of a right foot and a left shoe is to a left foot in a left shoe. The two salt molecules differ, in solubility in water for instance, just as their foot-shoe analogues are different from each other in ways more fundamental than subtle mirroring.

This is the beginning of a "resolution" of a mirror-image pair of molecules. Left and right feet are difficult to sort—especially if they are so tiny that you can't see them! Left feet in right shoes are easier to separate from left feet in left shoes (a bunch of conscripts encumbered with the mismatched shoes would march slower than a control group with the other). We may separate the molecules in the laboratory with ease—crystallize the D-L salt, leave the L-L salt in solution, then, after decanting we evaporate the solution of the latter in a separate beaker. The two substances now separated, we proceed to do the chemical analogue of taking the feet out of the shoes. This is a reaction that takes the salts back to the acids and bases that they came from. Separate beakers of D-L and L-L salts become separate flasks of D and L acids. Nothing is lost to a bug, if we do this neatly. (See the illustration on page 95.)

Such resolutions are practiced on an industrial scale today. And, as more and more pharmaceuticals must be marketed in one and just one enantiomeric form, people have devised clever ways of building molecules that are handed from scratch.[25]

VI

So one can separate handed molecules. But which is which? This question, which bothered the Jerusalem Talmud, Rashi, and Nahmanides, is metaphorical, even as it is absolutely essential to the functioning of a society. The chemical question is quite material (yet perhaps not so important). Remarkably, it was not answered until 1951.

There is no simple relationship between the direction in which a molecule rotates the plane of polarized light and the real geometrical arrangement of the atoms in that molecule. To put it another way, molecules that rotate the plane of polarized light to the right may or may not be "left-handed." Chemists have clever ways of relating spatial arrangements in one molecule to another molecule transformable into the first one by a sequence of chemical reactions. By 1951, in this way, they related *all* molecules to just one, called glyceraldehyde. But the form of glyceraldehyde that rotated the plane of polarized light to the right could have been the one chemists were

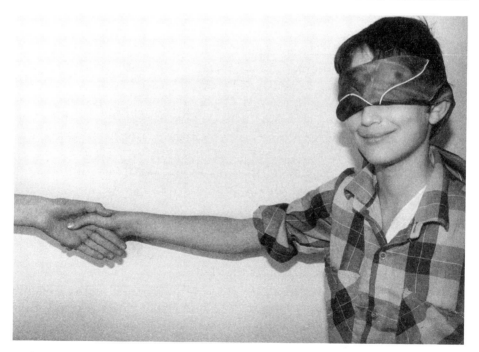

A blindfolded Michael Leibowitz differentiates between his sister's right and left hand; some chiral molecules are separated by a similar strategy. (Photograph by Kenneth L. Fischer.)

drawing. Or its mirror image. We knew we had a 50 percent chance of being right.

Something happened in 1951. In a very clever experiment the Dutch crystallographer and chemist Johannes Martin Bijvoet (pronounced "byfoot") determined for the first time the "absolute configuration" of a molecule.[26] We had to wait that long to find out that our guess about which way is right . . . had been right.

There is a deeper significance here, as an astute observer of science and literature, Stephen J. Weininger, observes:

> The passage from Rashi, in which he admonishes the reader to obey a judge who tells him that what he believes to be left is right, immediately (well, after a little reflection) evokes an analogy in chiral phenomena. The *appearance* of "rightness" or "leftness," as evidenced by the sign of optical rotation, is no guide to the actual "rightness" or "leftness" of a molecule in terms of its molecular structure. One may then bind Rashi and optical rotation together as being comments on different levels of truth and reality, one on the surface and the other deeper down.[27]

VII

Adriaen van der Werff was one of the many talented Dutch and Flemish painters of the seventeenth and eighteenth centuries. van der Werff was born in Rotterdam in 1659, studied with Eglon van der Neer, and was active in the city of his birth and in Germany until his death in 1722. In his lifetime, van der Werff was one of the most acclaimed artists of Europe, the court painter to the Palatinate Elector in Düsseldorf.

The illustration on page 97 shows van der Werff's *Jacob Blessing the Sons of Joseph*. And its mirror image. Now, a quick quiz: Which is the original, and which is reversed?

It helps to know the subject. The painting is Dutch. Alfred Bader, in his introduction to the exhibition "The Bible Through Dutch Eyes," in which this painting was shown, states:

> One of the remarkable aspects of life in seventeenth century Holland was the study of the Bible and the identification of Dutchmen with the People of the Book and their destiny. Never before had a Christian people studied the Bible so carefully. In many Dutch families the Bible was read daily, page by page, every morning. Events which are obscure to us became common household knowledge. Just as the

Adriaen van der Werff, *Jacob Blessing the Sons of Joseph*, Allen Memorial Art Museum, Oberlin College. One of these is the original, the other is a mirror image. Which is which?

> Jews appeared in the Bible as the champions of God and freedom, so Dutchmen looked on themselves as the latter-day Israel, and on their fight with Spain as a fight against tyranny, of good against evil.[28]

Adriaen van der Werff's theme is taken from Genesis, Chapter 48. Israel (Jacob) is about to die. He summons his son Joseph, and Joseph's two sons, Manasseh and Ephraim. The story continues:

> Noticing Joseph's sons, Israel asked, "Who are these?" And Joseph said to his father, "They are my sons, whom God has given me here." "Bring them up to me," he said, "that I may bless them." Now Israel's eyes were dim with age; he could not see. So [Joseph] brought them close to him, and he kissed them, and embraced them. And Israel said to Joseph, "I never expected to see you again, and here God has let me see your children as well."

Joseph then removed them from his knees, and bowed low with his face to the ground. Joseph took the two of them, Ephraim with his right hand—to Israel's left—and Manasseh with his left hand—to Israel's right—and brought them close to him. But Israel stretched out his right hand and laid it on Ephraim's head, though he was the younger, and his left hand on Manasseh's head—thus crossing his hands—although Manasseh was the first-born. And he blessed Joseph, saying,

> *The God in Whose ways my fathers Abraham and Isaac walked,*
> *The God Who has been my shepherd from my birth to this day—*
> *The angel who has redeemed me from all harm—*
> *Bless the lads.*
> *In them may my name be recalled,*
> *And the names of my fathers Abraham and Isaac,*
> *And may they be teeming multitudes upon the earth.*

When Joseph saw that his father was placing his right hand on Ephraim's head, he thought it wrong; so he took hold of his father's hand to move it from Ephraim's head to Manasseh's.

"Not so, Father," Joseph said to his father, "for the other is the first-born; place your right hand on his head." But his father objected, saying, "I know, my son, I know. He too shall become a people, and he too shall be great. Yet his younger brother shall be greater than he, and his offspring shall be plentiful enough for nations." So he blessed them that day, saying, "By you shall Israel invoke blessings, saying: God make you like Ephraim and Manasseh." Thus he put Ephraim before Manasseh.

Now it is clear which is the correct picture—van der Werff read his Bible carefully.[29]

But there is more. Heinrich Wölfflin, in a perceptive essay, "Right and Left in the Picture," has argued that quite aside from depiction of reality (blessings given with a right hand . . .) Western art shares a deeper pictorial code. A painting is not just a representation of reality, but is a constructed set of symbols, purposefully designed to be read in a certain way. Through an analysis of Raphael's *Sistine Madonna* and his cartoons for carpets in the Vatican, a Rembrandt landscape with three trees, and Janssen's *Woman Reading*, Wölfflin makes a case for a preferred entry point into a work of art (at the left), a favored mode of reading (rising along a left-to-right diagonal), and a concentration of forms and intensity of color at right.[30]

Look again at the mirror image of van der Werff's painting. Whereas one accepts the empty space among the unfurling curtains in the original (in which Jacob's right arm is outstretched) it looks terribly void, disturbingly so, in the mirror image (left arm outstretched). The latter's figures are cramped, hemmed in the left side of the picture; there is so much more visual space for them when they are in the right half. Joseph comes to peace with his father's intent countering of primogeniture in the original. In the mirror image he looks to us less accepting, more discomfited.

It will not have escaped the reader's notice that the biblical text is inordinately concerned with matters of right and left—the placement of the children by Joseph, his father's switch. Jacob in fact foresees the greatness of the descendants of Ephraim (among them Joshua and many kings of Israel). Jacob's reversal of the law of Israel is to be read in many ways—as prophetic insight to be sure, but perhaps also as a recognition that humanity must interfere in the workings of nature. And also as an echo of what happened to Jacob himself (the story of Jacob and Esau, and linking up to Cain and Abel, and Isaac and Ishmael, and Reuben and . . .). Curious, how left and right matter so much.

Not only do left and right matter, but they do so in the context of authority. There is Joseph, respectful of his father, seeking Jacob's blessing for his sons. Joseph sees Jacob place his hands in a way he, Joseph, perceives as wrong. He questions his father, reminds him, gently, of the law of the people, suggests that a mistake had been made. Jacob, just as gently and firmly, persists. Wills are expressed, actions taken, around an issue of precedence.

A final chemical note: van der Werff's painting has attracted technical interest because it uses an effective, stable, and inexpensive blue pigment, which has been identified as Prussian blue. We will meet this molecule, potassium ferroferricyanide, in Chapter 5, "The Flag That Came out of the Blue." Prussian blue was the first synthetic pigment. Just as for aniline dyes in textiles, it did not take long for artists to pick up a good pigment. Prussian blue was synthesized in 1704; van der Werff used it by 1722.[31]

VIII

We began this chapter by stating that carbon was tetrahedral, and hence derived the potential of chirality. We subverted the chronology a bit there. To restore it, there is another French-Dutch story to tell, and in it as well lurks the shadow of authority.

Pasteur, while he did go on, as we know, to other problems, remained fascinated by asymmetry. He experimented with the effect of magnets on crystallization, and he tried to reverse the sun's rays illuminating a plant, to see if this would induce synthesis of mirror image molecules. Pasteur speculated, without pursuing the idea in detail, that molecules were shaped like left- and right-handed helices, somehow steering the light waves.

The realization that a tetrahedral carbon atom could explain chirality and the existence of enantiomers had to wait for two young men, Jacobus Hendricus van't Hoff, and Joseph Achille Le Bel. [Please note the first names of van't Hoff and Le Bel, in relation to the previous section.] Though both had once worked in the same laboratory (that of one of the leaders of chemistry at the time, Wurtz), their discoveries were independent. In 1874, within a month of each other, when van't Hoff was 22 and Le Bel 27, they published papers written from very different perspectives, yet postulating the same source of chirality—a tetrahedral carbon center.[32]

van't Hoff's Dutch pamphlet, soon translated into French and German, was very directly (and provocatively for the period) entitled in French, *Chemistry in Space*. His presentation was more accessible than Le Bel's. What van't Hoff said was clear and convincing, and it was phrased in a pictorial language that is similar to the one we used at the beginning of this chapter. We reproduce on page 101 some of van't Hoff's models. Yet just because van't Hoff's thesis was so "geometrical" it predictably aroused the ire of those chemists who were loath to assign geometric significance to chemical formulas. For there was then no experimental technique to probe molecular structure.

One of these stalwarts of the old guard was Herrmann Kolbe, a great German chemist, the first to synthesize an organic molecule from inorganic reagents. Sadly, by 1874 Kolbe was in decline.[33] His criticism of van't Hoff (who by this time, 1876, was teaching at the Veterinary College in Utrecht) needs to be quoted *in extenso:*

Tetrahedral models used by J. H. van't Hoff.

In a recently published paper . . . I pointed out that one of the causes of the present-day retrogression of chemical research in Germany is the lack of general and, at the same time, fundamental chemical knowledge; under this lack no small number of our professors of chemistry are laboring, with great harm to the science. A consequence of this is the spread of the weed of the apparently scholarly and clever, but actually trivial and stupid Nature Philosophy, which was displaced fifty years ago by exact natural science, but which is now brought forth again, out of the storehouse harboring the errors of the human mind, by pseudoscientists who try to smuggle it, like a fashionably dressed and freshly rouged prostitute, into good society, where it does not belong.

Anyone to whom this concern seems exaggerated may read, if he is able to, the book by Messrs. van't Hoff and Herrmann on *The Arrangement of Atoms in Space*, which has recently appeared and which overflows with fantastic foolishness. I would ignore this book, as [I have] many others, if a reputable chemist had not taken it under his protection and warmly recommended it as an excellent accomplishment.

A Dr. J. H. van't Hoff, of the Veterinary School at Utrecht, has no liking, it seems, for exact chemical investigation. He has considered it more convenient to mount Pegasus (apparently borrowed from the Veterinary School) and to proclaim in his *La chimie dans l'espace* how the atoms appear to him to be arranged in space, when he is on the chemical Mt. Parnassus which he has reached by bold flight.

The prosaic chemical world had little liking for these hallucinations. Therefore, Dr. F. Herrmann, Assistant at the Agricultural Institute at Heidelberg, undertook to give them wider vogue by means of a German edition. This carries the title *"The Arrangement of Atoms in Space,* by Dr. J. H. van't Hoff; translated into German from the author's monograph by Dr. F. Herrmann, Assistant at the Agricultural Institute in Heidelberg; with a foreword by Dr. Johannes Wislicenus, Professor of Chemistry at the University of Würzburg. . . ."

It is not possible to criticize this work even half-way thoroughly because the play of fantasy therein dispenses completely and entirely with factual basis and is absolutely unintelligible to the sober scientist.

It is indicative of the present day, in which critics are few and hated, that two practically unknown chemists, one from a veterinary school and the other from an agricultural institute, judge with such assurance the most important problems of chemistry, which may well never be solved—in particular, the question of the spatial arrangement of atoms—and undertake their answer with such courage as to astonish the real scientists.

As I have said, I would have taken no notice of this work if Wislicenus had not inconceivably written a foreword for it and, not jokingly, but in complete seriousness, warmly recommended it as a worthwhile contribution—whereby many young inexperienced chemists might be misled into assigning some value to these shallow speculations. . . .

It is a sign of the times that the modern chemists feel themselves qualified and able to give an explanation for everything, and, when the results of experience are not sufficient, they seize upon supernatural explanations. Such treatment of scientific questions, which is not far removed from the belief in witches and from ghost-rapping, even Wislicenus considers to be admissible. . . .

Wislicenus thereby makes it clear that he has left the ranks of exact scientists and has gone over to the camp of the Nature Philosophers of ominous memory, who are only by a narrow "medium" separated from the spiritualists.[34]

Wow!

Kolbe's diatribe engenders a host of reactions, ranging from "It's a pity this great chemist is to be remembered by this polemic," through "This could be used as a compendium of tactics of calumny," and "Typical of these nineteenth century white guys with beards to invoke a prostitute out of nowhere," to "Too bad, they don't write 'em like that today!"

What interests us here is the multidimensional appeal to authority. Wislicenus's work appealed to van't Hoff—while he did not seek Wislicenus's authority to publish his work in Dutch or French, van't Hoff invokes the German chemist through an epigraph in his paper. Wislicenus, impressed by the very young Dutchman's work, sees to it that it is translated into German, and does append his *imprimatur* through a preface. The irascible Kolbe, aging and forgetting his own youth, invokes multiple mantles of authority as he rails against the *Naturphilosophen* (translated as Nature Philosophers, because Natural Philosophy has always had a good connotation in English), spiritualists, other scientists, and nearly everyone under the sun. And it's clear that Kolbe dislikes Wislicenus almost as much as he dislikes van't Hoff.[35]

Obviously authority matters in science, and not just for angry men unable to appreciate the startlingly new that turns out to be right—sorry!— correct.

IX

Diatribes also occur in Jewish religious texts. For example Nahmanides has this to say about his worthy predecessors Maimonides and Rashi, regarding their explanations for Moses' sin in striking the rock (Numbers 20:8–13): "These are the words of Maimonides, of blessed memory. He has added one more fatuous explanation to the other fatuous explanations!" The Hebrew actually invokes the famous opening of Ecclesiastes: "Vanity of vanities."

Right and left have not only metaphoric value in Jewish religious tradition, they have legal import as well.[36] Almost two dozen commandments can be affected by the handedness of a person. Among them: In which hand should one hold a palm frond and citron on the Feast of Tabernacles? In which hand should one hold the ritual cup when saying the blessing over wine? Which way should one lean during the Passover Seder meal? On which side does one blow the ram's horn sounded at the New Year? From which foot should a shoe be removed during the ceremony called *halitzah*[37]? On which doorpost does one affix a *mezuzah*? And even—which hand should one use for bathroom hygiene?

The questions that have generated the most discussion over the centuries concern *tefillin*, the black leather boxes containing biblical verses, attached with leather straps to the forehead and biceps. On which arm must a man place the phylacteries during morning prayers?

Chagall got it right, or rather left, in his portrayal, *Jew in Black and White*, in which a right-handed worshiper correctly placed the *tefillin* on his left arm. At least in one painting. In another depiction, it seems he decided the other arm was right (see the illustration below[38]).

Painting and etching by Marc Chagall. (*Left*): *Jew in Black and White*, Stiftung Karl und Jürg Im Obersteg; (*Right*): *The Grandfathers*, Sprengel Museum, Hannover. © 1997 Artists Rights Society (ARS), New York/ADAGP, Paris. Did Chagall intend to make one of them a lefty?

Despite the fact that the right side is given prominence in most precepts, for *tefillin* the left arm is prescribed. The Torah does not use the term right or left, but says (Exodus 13:16) "And it shall be for a sign upon thy hand" (thy hand= *yad-khah*). The unusual Hebrew spelling of *yad-khah* instead of the normal *yad-kha* is the basis for the rabbis seeing the concept of *yad-kehah* (the weaker hand) implied here. This in turn leads to the normative ruling that the left hand is signified.

But "weaker hand" is a relative term and this has led to a multiplicity of questions through the generations[39] with practical, anthropological, and philosophic implications.[40] There are two main schools—the absolutists who opine that the left should be used in most ambiguous circumstances, and the relativists, who use various criteria for determining which is the weaker hand (weaker vis-à-vis writing? when participating in sports? when using tools? when lifting?). Rabbi Yitzhak Luria, the Ari, of sixteenth-century Safed is the most extreme representative of the first school. According to him, one always places the *tefillin* on the left hand. Most authorities fall into the relativist category.

As an illustration of the process of determining the answers to these gray-area questions, consider the following query put to a prominent decisor of *halakhic* questions, Rabbi Yitzchak Yaakov Weiss (d. 1989), and his responsum:

> **Question:** A man while serving in the military, had his left arm tattooed, and in the place where the arm *tefillin* are put there is an image of a naked woman. The design cannot be removed. Now he has become an observant Jew and has begun to pray. He wants to know if he may place the *tefillin* on his left arm, on this image, or should he put it on his right arm?
>
> **Responsum:** In my opinion, he should put the *tefillin* on his left arm. Even if the image extends over the entire biceps, a large part of it can be kept covered continually, and he should expose only the area where the hand *tefillin* is placed in such a way that the entire image is not quite visible. He should have the *tefillin* made in the smallest size that is *halakhically* permitted. While reciting the blessing, he should keep the entire area covered. I have heard that it is possible to have a tattoo removed by an expert.[41]

X

darlin' I was left—right—Out of your arms
Oh, I was left—right—out of your charms.
My heart said eyes right! Right from the start
Now I am Left Right Out of Your Heart[42]

XI

There is no way to avoid the question of origins—whence our (macro) handedness, whence the (micro) handedness of molecules? Not all biochemical processes make one enantiomer, but most important molecules of life are handed—nineteen of twenty of the most common amino acids are of one type, so-called L (call it left-handed).[43] All sugars (and that includes not only the sweet kind, but also the sugar piece of nucleic acid backbones) are of the D type (call it right-handed, D for *dexter*). That amino acids are L, and sugars D, is true not just for us but for the biochemistry of every marvelous living organism.[44]

Why? Because in all organisms these universal building blocks are constructed by more complicated molecules, enzymes, which are also handed, like gloves. The instructions for the handedness of biochemical chiral molecules come down from the parents, encoded in chiral, helical, handed molecules.

So why do the enzymes and nucleic acids of all plants and animals exhibit that certain handedness, be it D or L? Because (so the accepted scenario goes) they descend through evolution from a common prebiotic chemistry.

In that primordial soup of the imagined beginning, the molecules pursued their nonappointed rounds—reacting to build complexity, falling apart. Take that chemical stone soup—in an ocean-size pot, with time on its hands. Take it to some point close to the essential stage in the development of life, where highly specialized molecules (call them protoenzymes) approach the ability to make more copies of themselves. These protoenzymes are likely to be handed, just as the real enzymes in life today.

Normally that soup produces equal numbers of left- and right-handed protoenzymes, their assembly aided by a catalyst, say the surface of a crystal. And that crystal happens to be one of those that Pasteur was lucky enough to find—no, not tartaric acid, but a crystal with similar properties of spontaneously separating into mirror-image forms. Now imagine a chance event in the soup. The catalyst crystals separate into left- and right-handed forms. In one place, call it A, by chance there is more left-handed catalyst, and it causes the formation, say, of a local excess of left-handed protoenzymes. Elsewhere in the soup, say at B, where the right-handed catalyst crystals are, there is a different microclimate and no protoenzyme is made. The result? A chance excess of left-handed protoenzyme molecules. Which multiply. Theoretical models show that it is perfectly possible that a slight excess of one enantiomer can blossom into dominance.

This is the aleatory, or "chance," mechanism of the origins of chirality. It exists in many variants, whose difference resides in the conditions of the first selective chemical act and in the placement of that act in the scenarios for the evolution of life.

How could one possibly prove or disprove this mechanism? Let us suggest an experiment that would convince some, though we're going to have to wait a while to perform it. Imagine we encounter a statistically significant sample (say ten) of extraterrestrial life forms. They are guaranteed to look like BEMs (bug-eyed monsters) because nature is a tinkerer.[45] But their chemistry will be our chemistry, adapted to their ambient conditions. If near half of these life forms have L-amino acids, and the other half have D-enantiomers, then we'll have a very strong plausibility argument for an aleatory beginning for all of us. (See the illustration on page 108.)

One variant of the chance hypothesis, which only pushes the "why" back in time, is that the chirality we know is not of earthly origin but was imported here (again, perhaps at a crucial evolutionary stage) by an enantiomer hitching a ride on, say, a meteor. Or a BEM.

William Bonner, who has worked long on this problem, said at a recent contentious meeting on "The Origins of Homochirality in Life":

> I spent 25 years looking for terrestrial mechanisms for homochirality [the exclusive existence of one enantiomer in biochemical molecules] and trying to experimentally investigate them and didn't find any supporting evidence. . . . Terrestrial explanations are impotent or nonviable.

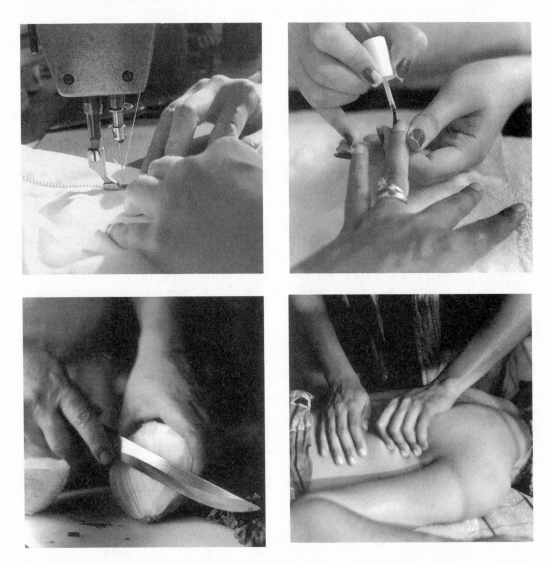

Hands of immigrant workers in Manhattan. (Photographs by Beatriz Da Costa.)

He further asserted that the theory that homochirality did not precede the origin of life was equivalent to: "believing in the tooth fairy or magic wands."[46]

The theory of extraterrestrial origins of handedness is sometimes called "panspermia." We haven't investigated if its proponents are primarily male.

There is another approach to the problem, which derives from an inherent handedness in some physical forces operating everywhere in the universe. Are you ready for some more physics?

The forces of the universe range widely in strength—they include the strong nuclear force, weak gravity, medium-strength electricity and magnetism (responsible for chemistry), and weak nuclear forces. The latter two have been combined in one unified theory by S. Weinberg, A. Salam, and S. L. Glashow. For a long time it was thought that all these forces were in a way even-handed, that they did not distinguish between left and right. Technically this was called "parity conservation."

From the 1920s one saw anomalies in the physical phenomenon of beta-radioactivity, which eventually led, in 1957, to C. N. Yang and T. D. Lee (both the fathers of superb chemists) to postulate "parity nonconservation," which was experimentally proven by C. S. Wu.

We now know that there *is* a basic handedness in nature; the electrons emitted in beta decay are inherently left-handed, the positrons right-handed. Nuclei are also handed, ever so slightly. Because there is an asymmetry to the "electroweak" force, there is an inherent difference in various properties, and in particular a difference in the energy of L- and D-enantiomers of a chiral molecules.

So there is a difference. But is it different enough to matter? Yes and no. The energy difference has been estimated for amino acids and sugars. It is a very, very small energy, even if it does favor in both cases the enantiomer dominant on earth. The energy differential is so small that as a result of it there would be in a reaction forming both molecules in a chemically nonhanded environment an excess of one molecule in about a hundred million billion[47] (1 in 10^{17}). Is that enough to matter? Here opinion divides. Some think yes, if that excess occurs under conditions of critical amplification. Some think no.

If our future experiment (those ten extraterrestrial forms of life) is ever done, and if all ten have D-sugars and L-amino acids, as we do, will you believe the underlying explanation of the electroweak force? Or will you invoke Creation?

XII

Interesting as it is, the question before us is not really that of left versus right, be it in science, religion, or art. No, the underlying problem is that of authority. How then does religion or science deal with authority? How do individual persons of faith and/or science react to imposed or accepted authority?

It is no accident that Jewish religion looks awfully authoritarian to an outsider. There is the commanding, peremptory God of the Torah. There is also this panoply of observance, whose externalities (work prohibitions on the Sabbath, the dietary laws, etc.) seem anomalous, even to many Jews, in this modern world.

A glance at a page of the Talmud (see the illustration on page 63), and this image of authoritarianism begins to evaporate, to be replaced by a sense of respect for intellectual authority. In the core text, the Mishnah and *Gemara*, we see opposing views as to observance laid out. So, for instance, there is a discussion on the correct way to light the Hanukah candles.[48] "The school of Shamai maintain: On the first day eight lights are lit and thereafter reduced [by one each day for eight days]. But the school of Hillel say: On the first day one is lit and thereafter they are progressively increased." Philosophical, logical, and historical reasons are given for each view. The view of the school of Hillel prevails (usually, as in this case, though not always); but both are said to be "the voice of Heaven."

It seems that questions of left and right are never far away. Further along in this discussion there is more disagreement: "And where is the Hanukah menorah placed? R. Akha, son of Raba said: On the right hand side [of the door]. R. Samuel of Difti said: On the left." And often, not always, a resolution: "And the law is, on the left, so that the Hanukah lamp shall be on the left and the *mezuzah* on the right."

The format of a page of Talmud today is a semiotic record of contentious debates and questioning by twenty-two centuries of intense scholars. Toward one margin we find the medieval commentary of Rashi, toward the other that of the generation of his grandsons. Both the central text and these commentaries contain references or citations; others adorn the periphery of the page. Still other commentaries are given as appendices to the tractate.

There is a clear sense of debate on absolutely every issue under the sun in these documents.[49] At the same time there is a tremendous respect communicated for the sages, the generations of religious scholars who have gone before. And from that balancing act of questioning and respect for tradition comes a tension that we think in many ways is similar to tensions in science.

The parameters that define the debate on authority arise out of certain classical passages in the Talmud. The first is the remarkable story of the encounter between Rabbi Eliezer ben Hyrkanus and a scholar whom we met earlier, Rabbi Yehoshua ben Hananiah.[50] Rabbi Eliezer declares a certain kind of oven impure. His fellow rabbis, led by Rabbi Yehoshua, disagree with him, fiercely so. Here is what transpires next:

> On that day, Rabbi Eliezer used all the arguments in the world, but they did not accept them from him. He said to them: "If the *halakhah* is in accordance with me, let the carob tree prove it." The carob tree was uprooted from its place one hundred cubits—and some say four hundred cubits. They said to him: "One does not bring proof from a carob tree." He then said to them: "If the *halakhah* is in accordance with me, let the channel of water prove it." The channel of water turned backward. They said to him: "One does not bring proof from a channel of water." He then said to them: "If the *halakhah* is in accordance with me, let the walls of the House of Study prove it." The walls of the House of Study leaned to fall. Rabbi Yehoshua rebuked the [walls], and said to them: "If talmudic sages argue with one another about the *halakhah*, what affair is it of yours?" They did not fall, out of respect for Rabbi Yehoshua; but they did not straighten, out of respect for Rabbi Eliezer, and they still remain leaning. He then said to them: "If the *halakhah* is in accordance with me, let it be proved from Heaven." A heavenly voice went forth and said: "Why are you disputing with Rabbi Eliezer, for the *halakhah* is in accordance with him everywhere?" Rabbi Yehoshua rose to his feet and said: "It is not in heaven."

Generations later, the Talmud relates:

> Rabbi Natan met Elijah and said to him: "What did the Holy One blessed be He, do at that time? He said to him: He smiled and said 'My sons have defeated Me, My sons have defeated Me.'"[51]

"For it is not in heaven!"[52] The religious laws of observance, of everyday life and service to God, are to be derived by human beings on earth. This is a truly remarkable dictate.

But those steeped in learning are to be listened to: "'You shall fear the Lord your God'—this includes Torah scholars" it says in the Talmud.[53] This is not just self-serving propaganda from the Babylonian groves of academe, but a reflection of a sincere and abiding respect for scholars throughout Jewish history. There is no shortage of examples to this day of the sacrifices that a family (yes, borne mostly by wives . . .) will make to support a gifted religious scholar.

For over two millennia the basic texts of the Jewish religion have been painstakingly commented on and debated. The record of those discussions was first passed on orally, then written, then printed. It is already entering the electronic age, to wit Bar Ilan University's database of rabbinic responsa. Despite disagreement, the prevailing tone is one of respect for books and the people who study them.

Should one follow the older, the more venerated scholars? Interestingly the answer given is usually "no," with no disrespect to the earlier sages implied. The religious law follows the opinion of the later scholars. To quote the Rosh (d. 1327):

> The statements of the later scholars carry primary authority
> because they knew the reasoning of the earlier scholars as well as
> their own, and took it into consideration in making the decision.[54]

So, the Torah and the *halakhah* are not in heaven. It is a living Torah, and resides in the inspired minds and mouths and pens of fallible men and women, who may—no, are certain to—disagree with each other. How is one to deal with that disagreement?

First with humility and the acceptance of the fact that there is bound to be dissent. As Nahmanides put it:

> And you who look into my book, do not think that all of my replies
> are all (in my eyes) convincing, compelling you to agree with them
> despite your own objections, with the result that you will boast-
> fully refute one of them, or coerce yourself to dismiss my proofs.
> Such is not the case. For anyone who knows the way of our
> Talmud knows that there are no final proofs in the disputes among
> its commentators, nor are there generally absolute refutations. For
> in this science there is no clear proof, as is the case with algebra
> and astronomy.[55]

And second, the religious scholar facing the issue of disagreement is clearly enjoined with the right and responsibility to dissent, if there be reason to do so, in the face of authority, as this sensitive passage in the Talmud affirms:

> Rava said to Rav Papa and to Ravina b. R. Joshua: Should a decision of mine come before you, and you notice a flaw in it, do not tear it up until you have brought it to me. If I have an argument (i.e. in reply to the objection), I will tell you. If not, I will retract. After I die, neither tear it up nor learn from it. Do not tear it up, for perhaps if I were still alive I would have resolved the difficulty. Do not learn from it, because a judge must rely upon his own opinion.[56]

The admonition to come and talk is wonderful, and while it is a very personal plea, it reflects the general respect for debate and the dialectic. But in the end Rava is affirming here the right, no, the *responsibility*, of a scholar to question what has gone before.[57]

Scholars, treating each other with respect, recognizing the value of what each is trying to say in the light of their fathers, their teachers. . . . Trying to find, each in his or her own way, inspired by heaven, the Torah that is not in heaven. But they are human, after all, and each sees the world in his or her own terms.

There is a good example of this in a contemporary political/religious struggle in Israel. Recent events in the Middle East included an accommodation of the State of Israel with the Palestinians, a substantive part of which is the withdrawal of Israelis from the territories occupied after the 1967 war: Gaza and the West Bank (Judea and Samaria). The policy of withdrawal does not meet with approval, to put it mildly, of a significant part of the population of Israel. The issue is highly divisive.

In 1993 the rabbinic heads of some yeshivah-army programs[58] issued a declaration, urging their students (and all observant Jewish soldiers) to disobey orders if they were to withdraw or aid withdrawal. Heads of the other yeshiva-army programs abstained from such a call or provided *halakhic* counter-arguments.[59]

A tradition always subject to contentious, emotionally charged debates—this is what the striking passages quoted above imply. To a nonreligious outsider, Jewish observance looks like an awfully restricted, authoritarian life, with no room for debate. But debate is probably the one common thread that runs in Jewish tradition over the warp of the centuries and the woof of geography.[60]

Handedness is a quality not only of hands and molecules. A sculpture in Rhode Island. (Photograph by István and Magdolna Hargittai.)

XIII

Sam Orbaum asks:[61, 62]

> If the Left were right and the Right were left out, would it be right for the Right to be left out outright and right to write off the Left right off?

☐ Write   ☐ Right   ☐ Riot

XIV

Maybe it's not that different from science. Here we have a marvelous construction dedicated to gaining knowledge, the persistent teasing out through experiment of the workings of nature, and the making of so much new. This is accomplished by ever-skeptical researchers, communicating their results freely. Everything is open, everything is subject to debate. One group reports it has found the gene predisposing women to a certain type of breast cancer; another group disputes the finding in an issue of *Science* a month later. It's all done in gentlemanly (if neutered) tones, reporting the facts and nothing but the facts.

But how does it look to the beginning student in that introductory chemistry course one of us teaches? Professor H. tells the students that in a series of experiments in the nineteenth century people found that for the reaction

$$aA + bB \rightleftharpoons cC$$

at equilibrium the ratio of concentrations raised to the power of their "stoichiometric coefficients" (a, b, c) is constant:

$$[C]^c/[A]^a[B]^b = K$$

He does *not* tell them of the struggle that went on to prove this "law of mass action," the many misleading results that seemed to invalidate the law, the fact that one should really use "activities" and not concentrations. No time for that. Nor do the students want history, only what is on the medical school entrance exams.

And after Professor H. tells them all this, he seals the knowledge in their minds by giving them an examination that asks them to manipulate this "equilibrium constant" expression. Professor H is then happy if the students show him their control of the concept by an 80 percent performance on the exam.

The student who stops to ask just how good the experimental proofs for the equation are, or who has the audacity to try another equation, will probably not do well in the course, and might get a reputation as a nuisance. There is so much knowledge that is certain by today, so much learned by the giants on whose shoulders we stand! Who would question Newton or Maxwell? To a reflective beginning student, the introductory science course (for that matter nearly every course up to graduate school) has a very authoritarian feel to it. It's a real educational problem for our profession; the vaunted openness of science is not obvious to our apprentices.

Questions of authority are inherent in the fabric of modern science. Take the system of referencing previous work through footnotes. While there may indeed be several reasons for referencing a paper, primary among them is a recognition of (and trust in) a previous measurement or synthesis, for the purpose of using it in the current work. The edifice builds on what has gone before; the paper without references is a rarity.

This building on the shoulders of giants does not always proceed in a gentlemanly fashion, as this exchange between two of the great chemists of this century attests:

> ...but to be perfectly candid I think there is a chance that the casual reader may make a mistake which I am sure you would be the last to encourage. He might think that you were proposing a theory which in some essential respects differed from my own, or one which was based upon some vague suggestions of mine which had not been carefully thought out.
>
> Lewis to Langmuir

> In the first place we must realize that no one can or should have a proprietary right in a theory for all time.... Strictly speaking if the originators of the theory must be mentioned with it I think the theory should be called the Thomson-Stark-Rutherford-Bohr-Parson-Kossel-Lewis-Langmuir theory.
>
> Langmuir to Lewis[63]

There is substantive similarity between the methodology of the Jewish Oral Law (Talmud, commentaries, responsa) and science in that they share a commitment to close analysis and a tradition of citation. There is also a great difference between scientific patterns of citation, and those of Jewish religious discourse. In the rabbinical responsa, the opinion of the later decisors/scholars is normative. But the older, indeed the very old, decisions are *always* cited, with great veneration. In science, the cult of the new reigns. Seldom is a citation more than twenty years old.[64] It could be the pace of advance; we think it's in part a culture of valuation of the new. The Oedipal urge is stronger in science.

Whence does authority derive in science? From one's achievements, published, scrutinized, and found to be of value by one's peers.[65] Note the components—all necessary—that (a) there be a body of work done by an individual, (b) that the work be exposed to the public, and (c) that the community find it of value.

The work must be there, for without an ouevre there can be no reputation. The achievement need not be voluminous—the community is well able to recognize quality, even when it is dispensed in small doses. R. B. Woodward, the great synthetic organic chemist of this century, published just over a hundred papers. Ken Wilson, one of the foremost physicists of our time, publishes relatively little. We've seen colleagues promoted to tenure on the basis of four good papers, and we've seen colleagues not promoted who have written sixty routine ones. So much for "publish or perish"!

But publish one must, because science, like art, reaches out to an audience. It is about people communicating insights to other people. There is a difference between poetry and therapy. There are many motives for writing in science—to teach, to understand, to claim priority. Whatever reservations one has about the ossified form of communication that is the scientific article (and one of us has many problems with it[66]), one sees that the most effective part of the system of modern science is free exchange of information through publication. There are good commercial reasons for not publishing or for publishing in that curious way of the patent, but the advancement of reliable knowledge and the advancement of the individual depend on the achievement being seen.

Incidentally, unlike literature, it is trivially easy to publish in science. The world's best chemistry journal has an acceptance rate of 60 percent for full articles, 35 percent for brief communications of important findings. Fall a rank below (yes, there is a pecking order!) to an average-to-good journal, and the acceptance rate rises to over 90 percent. For poetry, the corresponding quality magazine will have an acceptance rate of 1 to 2 percent.

The community of science is nicely addicted to reading, or at least compulsively scanning, its own literature. It is true that some discoveries manage to be ignored. In a minority of cases it is through the author's own fault—the writing is obscure or the discovery is put in the wrong theoretical context. Most of the time it is the community that is at fault. When we say "a discovery is ahead of its time," we really mean that the scientific public was, for good or bad reasons, unwilling to consider it. It did not fit current paradigms or one did not see how to prove it or disprove it. Or the work may have failed Ockham's Razor arguments (yet been true!).[67]

But most experimental findings that are novel, or most theories that really explain rather than restate, are not ignored. They may be resisted if they argue with accepted or fashionable ideas, especially if those ideas are held by leaders in the field (the "great of their generation"). All the better. Fueled by the desire to prove those upstarts *wrong*, not right, people will replicate important experimental findings and test the consequences of wild theories. The reason science "works" is that it is a wonderful mechanism of coupling curiosity and personal ambition in a framework that yields something much greater than its parts—the cogs of the machine, the fallible, if smart, people who labor at and rejoice in accumulating knowledge.

Parenthetically, this is why fraud is truly unimportant in science—not because the idea of science is a search for truth, but because the process of science is so self-correcting. The psychology of falsifiers of data drives them to claim discoveries of import, not trivia. And there is something in the nature of science that makes people put to the test claimed discoveries not of trivia—for who has time for that—but that which is important or unexpected. The week the high-temperature superconductors were made, there were furnaces stoked making those ceramics around the world. The same furnaces have failed to replicate the claims of good people to have (up until now) made room-temperature superconductors.[68,69]

XV

There is an interesting religious counterpart to the problem of publication and peer review. Two of the great religious scholars of their time—of our time—were the late Rabbi Joseph B. Soloveitchik and Rabbi Moshe Feinstein, both from Europe originally, both living much of their lives in New York City. Though first cousins, they were quite different personalities; their attitudes to observance in modern life differed, and their views appealed to different segments of the observant community.

Rabbi Feinstein wrote voluminously and his responsa and general works were widely published. Through the cogency of his reasoning, and the wide dissemination of his work, he became a guide with authority. Rabbi Soloveitchik's intellectual and psychological depth was legendary. But Rabbi Soloveitchik, to quote Michael S. Berger:

> . . . as is well known, was reluctant to publish some of the decisions he has given. This approach has its merits, based on the concern of abuse and misuse of the decision once it is commended to the Jewish *reshut ha-rabbim* (public domain). If one wants to control the application of a decision, it must be confined to only those who asked the question, and who were trusted not to mishandle the decision. But the down-

side of such a strategy is the inability for others to review the argument(s) used to arrive at the conclusion, hence essentially excluding it as an opinion which may be considered by others when the subject, or a related issue, arises. It undermines the authoritativeness of the Rav's [Soloveitchik's] decisions (except to those who have heard the answer from him personally) in that others cannot even be sure whether the Rav continued to endorse the position attributed to him. Only those who came directly within the Rav's orbit (or that of his talmidim [students, disciples], will be able to consider him as authority. In contrast, Rav Moshe's [Feinstein] decisions, whether arguable or not, are available for all to see, to examine, and to analyze.[70]

Actually, these two men represent two streams in Jewish religious thought throughout the ages, two options available to all. Rabbi Shalom Carmy puts the choice in the following perceptive way:

> The question of right and left in *Halakhah* can be viewed in two very different ways. You may hold that the correct view must subjugate itself to the incorrect view for political reasons or you may hold that the court's decision in fact determines truth. Developing the second option you may hold that the court's performative function turns falsehood into truth: R. Yehoshua was right about the calendar, but R. Gamaliel had the authority to determine the calendar, and it was this determination that is valid. But you may look at the authority of the court from a different perspective: both rulings may be implicit in the Torah, each one is true, *Halakhic* decision-making means precipitating one of the legitimate possibilities.[71]

Rav Moshe Feinstein chose to be a decider, a *posek*. Actually it's not so simple. As Rabbi Feinstein himself said:

> You can't wake up in the morning and decide you're an expert on answers. If people see that one answer is good, gradually you will be accepted.[72]

Note the significance of evaluation of the work by the community. Not that different from science. . . .

Right- and left-handed snails on Borneo. (Drawing by Jaap J. Vermeulen.)

Rav Soloveitchik chose to be a teacher. Cyril Domb, a distinguished physicist who listened to Soloveitchik's lectures recalls a remark by him in the context of his being considered for the Chief Rabbinate of Israel:

> I would be useless as a Chief Rabbi. I am not a *posek*. I give a decision one day, and the following day I discover 100 reasons against it. My function is to be a *Melamed* [a school teacher].[73]

Soloveitchik's vocation

> was not to choose among the options offered by the classic *halakhic* literature, but to work out the deepest analysis and appreciation of each view.[74]

XVI

If you think that telling left and right apart has become more difficult after these scientific, religious, and artistic perambulations, consider the following: Snail shells can be right-handed (dextral) or left-handed (sinistral), describing their coiling.

But that's just the beginning. Edmund Gittenberger writes:

> . . . Until now, four kinds of snail-shell coiling have been distinguished, because (although this is rare) the soft body parts can be coiled in the opposite direction to the shell: dextral shells containing sinistral bodies and sinistral shells containing dextral bodies are known. Indeed, things are more complicated still, because the larval shell (which is often retained at the tip of the adult shell) and adult shell can have different hands.[75]

Recently J. J. Vermeulen has described thirty-six snail species on Borneo, 1–3 mm in size with a further twist (illustrated on facing page). They "start out" dextral or sinistral but their shells have a final whirl that is exactly opposite![76]

After reading this chapter the reader may legitimately feel like one of those poor souls of Nineveh, which the Book of Jonah[77] describes as "that great city, in which there are more than a hundred and twenty thousand persons who do not yet know their right hand from their left. . . ."

Bitter Waters Run Sweet

#1
Date: 11 Shevat (19 Jan)
From: Shira Leibowitz Schmidt (shiras@netvision.net.il)
Subject: The bitter waters of Marah

   I would like to ask the help of the Bibl-e-mail
threaded news-group in puzzling out a passage in Exodus:
How exactly did Moses sweeten the water that was too bitter for the children of Israel to drink?
   The episode comes on the heels of the crossing of the
Red Sea. After surviving the surfeit of water that drowned
their Egyptian pursuers, the Israelites then had to survive a dearth of potable water in the parched desert that
faced them. The events are described in Exodus:

> Moses led Israel from the Red Sea out into the
> wilderness of Shur. For three days they traveled
> through the wilderness without finding water. They
> came to Marah, but could not drink the waters of
> Marah because they were bitter; that is why the
> place was called Marah [Hebrew for "bitter"]. The
> people complained to Moses and asked, "What are we
> to drink?" Moses cried to the Lord, and the Lord
> showed him a tree[1] which he threw into the water,
> and then the water became sweet.[2]

   How do you interpret the last line? What really happened at Marah?

#2
Date: 20 Jan
From: Professor Yoram Zweifler
(yzweifler@immer.sicher.ac.il)
Subject: Marah

The Bible, as usual, is imprecise. I'll constrain my desire to take the chronicler of Exodus to task for his poor reporting (after all, he didn't have the benefit of the rigorous laboratory course at the University of Munich that I endured just before World War II). Let's take the meager clues the text provides and see if we can decipher the science on which this supposed miracle is based.

What could have made the waters of Marah taste bitter? I don't think there was any quinine (a very bitter tasting New World plant) for thousands of miles around. Perhaps the water was salty or brackish. Maybe it was "sweet" water, but polluted (before chemical industry!) by some local biological overburden of plant growth. Could such water taste bitter? Perhaps there were some mineral deposits nearby, and Lake Marah leached some of them, rendering the water bitter.

As poor scientific observers as the Israelites were (and my present students are no better), I'd doubt they'd call salty water bitter. So I'm inclined to think the water was polluted and bitter due to some mineral leaching. We need to look for such bodies of water in the northern Sinai.

It is clear that Moses then put into the waters of Marah a log that contained a naturally occurring chemical that in some way removed the source of the bitterness. To speculate at this point on precisely what happened (and was so imprecisely reported) would be unscientific.

#3
Date: Tu B'Shvat
From: Ira T. B. Liever (ir8@2xs.com)
Subject: The bitter waters of Marah

Who cares, Mrs. Schmidt, what tree it was that Moses threw into the water? As ibn Ezra said a thousand years ago, "We don't know what tree, only that it was *pele* (something wondrous)."[3] The miracle was a sign of the compact between G-d and His people, one of many, many such miracles.

#4
Date: 23 Jan (15 Shevat)
From: Andy Goldfinger
(andy_goldfinger@spacemail.jhuapl.edu)[4]
Subject: Bitter, bitter tastes

Your inquiry regarding the bitter waters of Marah is of importance to me. I am experiencing a medical problem in which I get "attacks" of bad taste in my mouth that are so intense that I often become disabled. (The taste varies in quality, but sometimes it is extremely bitter or metallic.) So far, the doctors have not been able to determine the cause.

While reading this week's Bible portion [Chapters 15 and 16 of Exodus], I was taken by both the bitter waters and the taste of manna. . . .[5] Is there some natural effect of the tree put into the bitter waters? Might I be able to use this to somehow cure my condition?

I would appreciate it if you would share with me anything you learn through the Internet or through any other sources. Thanks in advance for your help.

All the best—Andy Goldfinger

#5
Date: 23 Jan
From: Professor Yoram Zweifler
(yzweifler@immer.sicher.ac.il)
Subject: Ignorance and bitter waters

Mr. Liever is representative of the unenlightened, superstitious, unthinking crowd that has ruined the Zionist dream! I don't know (yet) the tree that Moses threw into the water. But I'm sure it was no miracle—there is certainly a good scientific explanation for what happened.

#6
Date: 16 Shevat (24 Jan)
From: Ira T. B. Liever (ir8@2xs.com)
Subject: Arrogance and bitter waters

After reading the message from Professor Yoram Zweifler, all I can say is Heaven preserve us from "*yekke*" professors![6]

Trying to find a scientific explanation is barking up the wrong tree. I propose we ignore the scientists and see what the sages say. One of the earliest explanations I know is a *midrash* that describes the episode at Marah not only as a miracle, but as a double-header, a miracle within a miracle:

> Humans harm with a knife and heal with a bandage. Not so the ways of G-d who harms and heals with the same instrument. So when they came to Marah and could not drink, Moses thought that G-d would tell him to toss some honey or dates into the water to sweeten it. But look what is written in Exodus: "Moses cried out to the L-rd and the L-rd showed him a tree." It is as if G-d said to Moses, "My ways are not the ways of man. Now you have to learn My ways."[7]

I couldn't say it any better; the *midrash* disabuses us of any notion that Marah might have been a miracle wrought by natural means.

#7
Date: 18 Shevat (26 Jan)
From: E. Z. Going (ezg@hal.oxbr.ac.uk)
Subject: Biblical botany and Marah

Let me log into the discussion about the identity of the tree Moses used. The rabbis must have been quite familiar with the botanical characteristics of various trees. Another *midrash* comments:

> Rabbi Shimon ben Gamaliel said: "Come and see how the Holy One's methods are more wondrous than those of mortal humans. Flesh and blood use something sweet as an antidote to bitterness. But the Holy One sweetens the bitter with something else that is bitter!" [As if] God said to Moses: "Since these are bitter waters, I will sweeten them with something bitter." How? He puts that which causes injury into the injured thing and so makes a miracle.
> And which tree was it? Rabbi Yehoshua says it was the willow, Rabbi Eleazar of Modiin says it was the olive (since there is nothing more bitter than

olive), Rabbi Yehoshua ben Korha says it was the oleander, Rabbi Nathan says it was the katros [cedar], and others say it was the fig or pomegranate.[8]

Nogah Hareuveni[9] has pointed out that the sages in this passage went beyond looking for a plant that could sweeten bitter water. They looked for a plant that grew near water, to be sure. But also driven by a "theory" of miraculous healing by like canceling like, they looked for trees that in some way were bitter. The oleander is such a tree, not only bitter but poisonous.[10] And anyone who has tasted a fresh olive off a tree knows that the fruit must somehow be processed before eaten. Olives contain a bitter chemical, called a glucoside, oleuropein.[11]

This idea of healing a disease by treating it with something similar is an ancient medical theory that has an interesting modern counterpart in vaccination. One infects an organism with a related or modified form of a virus so as to elicit an immune response that will protect against the real virus, if it attacks. There are also some compounds used in homeopathic medicine whose action is based on this concept.

#8
Date: 26 Jan
From: Professor Yoram Zweifler
(yzweifler@immer.sicher.ac.il)
Subject: Stupidity and Marah

There is no end to the tribulations of Israel! I'll ignore the nasty comments of Mr. Liever on my venerable European antecedents. But the *midrash* he quotes is an example of the outlandish imagination of biblical commentators. To think that throwing sweet organic matter such as dates into bitter water will make it sweet! All that one will get is decomposing, rotten fruit, and bacteriological contamination. . . .

Mr. E. Z. Going I thought I could talk to, for even if he quotes another farfetched *midrash*, at least he has done some reading on the flora of Israel, and begun to think about a plant with the appropriate qualities. But then he spoils it with a reference to that quackery, homeopathy!

#9
Date: 27 Jan
From: Roald Hoffmann (rh34@cornell.edu)
Subject: Commentary on bitter waters

I'm just a newbie, a theoretical chemist with no experience of bitter waters other than tonic water. I'm not
observant, but respectful of the tradition, and if I may,
I'd like to mediate between Professor Zweifler on the one
hand and Ira T. B. Liever and E. Z. Going on the other.

That the author of the *midrash* cited by Mr. Liever has
some simplistic water-purification ideas is easily for-
given. More interesting to me is his last sentence, "God
said to Moses, 'My ways are not the ways of man. Now you
have to learn my ways.'" This direct quotation of the
Almighty is strong. However, there is no biblical or talmudic authority for God speaking thus, and I take it as an
interpretation by the author of the *midrash*.

More than an interpretation, it is the staking out of
an ideological position. The position is a strong paraphrase of the views of Messrs. Liever and Going. The
midrash must be reacting to a view that secular knowledge
(no science as such then . . .) *is* relevant to the life of
a religious Jew. This *midrash* says that not only doesn't
it matter what science says, but that the ways of God are
fundamentally different from the ways of nature.

I don't accept that. And I'm not the only one. The
alternate choices of trees by Rabbis Yehoshua and Eleazar
of Modiin, and Yehoshua ben Korha, and Nathan are clear
testimony to the fact that the imperative to understand
the ways of God through nature is ancient and essential to
Jewish *religious* study, not to speak of everyday life.

#10
Date: 28 Jan
From: Molly Saltzwasser (tia@sopa.de.pollo.org)
Subject: Bitter or salty

In response to your Bibl-e-mail discussion about bitter
waters: A week ago at the Sabbath dinner table Torah discussion on Friday night, the bitter waters were briefly
touched on. I have always assumed these waters were salty
like seawater. My hydrologist daughter explained that the

water may have been brackish, as is our water here in Albuquerque.

My mother-in-law (she lives with us) then astutely pointed out that when her chicken soup comes out too salty, she adds potatoes. Which absorb the salt due to what? She said she learned it from her mother.

At this point my educated daughter opined that maybe it has to do with the fibrous nature of the vegetable. She hypothesized that if the tree were fibrous as potatoes are, then the wood might be capable of absorbing salts in slightly brackish water, making that water potable.

My mother-in-law said, "For this we sent her to Harvard?"

#11
Date: 20 Shevat (28 Jan)
From: Shira Leibowitz Schmidt (shiras@netvision.net.il)
Subject: Thanks for discussion

I appreciate all your inspired help, Bibl-e-mail friends, in getting me into deeper waters on bitter waters!

I've found something myself. Apparently not all sages agreed that there was a double miracle of bitter sweetening bitter. Hizkuni[12] points out the obvious, that God could have sweetened the water *sans* tree if He so desired. But it is His wont to perform miracles within the framework of the laws of nature, by adding a sweetening chemical into the bitter. Therefore I find nothing objectionable in ancient sages' or modern scientists' speculating about the natural mechanism that might have been invoked and shown to Moses.

Could I raise the wider question of miracles? I know this is a vast subject.[13] From what I read there was some ambivalence in rabbinic thought about miracles abrogating the laws of nature post Creation. For example, in the Talmud, tractate Avot 5:8 it says that some miracles were features of *the* miracle—Creation. Among the ten things created on the Sabbath eve at twilight were the mouth of the earth which engulfed Korah and his fellow conspirators (Num. 16:32), the rainbow, manna, and Moses' staff. Thus miraculous exceptions as well as nature's rules were built into Creation.

#12
Date: 28 Jan
From: E. Z. Going (ezg@hal.oxbr.ac.uk)
Subject: Miracles

A propos miracles, I found an interesting comment from one of my countrymen, George Bernard Shaw:

> When we refuse to believe in the miracles of reli-
> gion for no better reason fundamentally than that
> we are no longer in the humor for them we refill
> our minds with the miracles of science, most of
> which the authors of the Bible would have refused
> to believe. The humans who have lost their simple
> childish faith in a flat earth and in Joshua's feat
> of stopping the sun until he had finished his bat-
> tle with the Amalekites, find no difficulty in
> swallowing an expanding boomerang universe. . . .
> Religion is the mother of skepticism. Science is
> the mother of credulity. There is nothing that
> people will not believe nowadays if only it be pre-
> sented to them as science, and nothing that they
> will not disbelieve if it be presented to them as
> religion. I myself began like that; and I am ending
> by receiving every scientific statement with dour
> suspicion whilst giving very respectful considera-
> tion to the inspiration and revelations of the
> prophets and poets.[14]

#13
Date: 29 Jan
From: Perennial K. Vetch (peryk@sr3.catnet.org)
Subject: Some things never change

Jews, Jews—what a people! One person asks you a simple
question, about bitter waters. So what do you get? Three
opinions from authorities who couldn't agree two thousand
years ago, interpreted by four people who won't listen to
one another today. Also a flaming superrationalist
scoffer, someone with a bitter taste in his mouth and a
mother-in-law who knows how to sweeten brackish chicken
soup. And off we go into botany, geography, and miracles.
Can't you people stick to one subject? I think it's time
to split up our news-group. . . .

#14
Date: 22 Shevat (30 Jan)
From: Bracha Alhanissim (ba@ntb.desal.il)
Subject: Entebbe

Whether Marah was a miracle or occurred through natural means (even if Moses were instructed in chemistry by G-d) makes an essential difference to an observant Jew, to me. For if it were a miracle, then those who experienced the miracle and their children would be obliged to (and very much want to) make a benediction when they revisited the site.

It is a *mitzvah* to pronounce a blessing when the occasion arises—be it on eating brownies, seeing a rainbow, or meeting a wise scholar. Or visiting the site of a miracle. The formulation is "Blessed are You, O L-rd our G-d, King of the universe, Who performed a miracle on my behalf at this place." The *Shulhan Arukh*, an influential code of Jewish law, then makes an essential definition, as follows:

> . . . one should not make a blessing for a miracle unless the miracle involved a departure from the world's natural course. However, for a miracle that was in keeping with the world's natural course and was a normal development (for example, if thieves came at night and one was in danger, but was saved from them . . .) one is not obliged to make a blessing.[15]

Lest this be thought obscure, one might look at the controversy unleashed in Jewish orthodoxy by the question of blessing after Entebbe. You recall the dramatic rescue by Israeli soldiers led by Yonatan Netanyahu of a hijacked Air France jet at Entebbe on the Fourth of July 1976. The religious problem was whether it was appropriate (for hostages or their children, or rescuers) to pronounce just such a blessing for this miracle.

Various religious authorities split along predictable ideological lines (most orthodox groups support the State of Israel, even as they might disagree with policies of the party in power; but they differ vehemently in how they view a secularist Jewish state vis-à-vis divine providence). Several rabbis published (some in paid advertisements) their decisions to endorse or deny the validity of the miraculousness of the rescue.[16]

I was there, not at Marah, but at Entebbe! I was one of
the rescued passengers. Now, two decades later, I have an
impending sales trip (I'm in the business of building
desalination plants, so maybe I should have been at Marah)
which necessitates a stopover at Entebbe. My rabbi has
advised me that, while I must be thankful for deliverance,
I shouldn't confuse a real miracle with a dazzling rescue
executed meticulously. Therefore I should refrain from
reciting the blessing.

#15
Date: 31 Jan
From: Monsignor Aquadulce (aquadulce@x.pius.va)
Subject: Miracles in Catholicism

    May I tell my Jewish friends something about the expe-
rience of the Catholic Church with miracles? In our tradition, there has been an unabating flow of miracles. And we
have developed institutionalized ways of evaluating the
miraculous, ways that may be of interest to you.
    From the beginning of Christianity there have been
several distinguishable streams of miraculous behavior.
Just as in post-biblical Jewish tradition, curiously few
of these are caused by God directly, even if they are
generally and appropriately attributed to Divine
Providence. Most miracles are the result of prayer to
various intercessionary figures, such as Mary, the
saints, and, interestingly, men and women who have not
(yet) been recognized as saints of the church. The
importance of these intercessionary persons has to me a
curious parallel to the role of the *rebbe,* the charismatic leader of *hasidic* sects in Judaism. An important
difference is that the *rebbe* lives, while the Catholic
figures with similar powers to intercede with God on the
behalf of a human being are almost without exception
deceased.
    There are also several archetypical miraculous occur-
rences in Catholic tradition, such as the *stigmata.* Or the
stream of apparitions, mostly of Mary, in clouds, sheets,
stones, a great range of natural objects. Or the inexplicable (from a scientific perspective) behavior by relics or
images of the saints, for instance a painting of the
Virgin which weeps on a certain date.

A long time ago the Church recognized that sincere yet credulous believers may respond excessively to what they believe are signs from God. And believers may be deceived or manipulated by human concocters of "miracles," some of whom, sad to say, may reside even in the bosom of the Church. For much religious and secular influence, therefore, power, even money, may accrue from supposed miracles. So the Church, with the aid of modern science, has been very careful to probe whether a supposed miracle in fact exceeds the limitations of the natural. Thus there is an International Medical Committee at the great and truly miraculous shrine of the Virgin Mary at Lourdes. This committee reviews various claims of miracles occasioned by the Virgin at Lourdes and has certified the validity of some. But the committee is conservative; in 1948, 83 claimed cures were investigated, only 9 certified. There has been no claim validated since 1989.

The most elaborate mechanism for verifying the miraculous event in the Catholic Church is set into motion by nominations for beatification and canonization of saints. In early Christianity a popular cult of veneration of any figure created a saint. Some of those early saints, associated to this day with professions or certain causes, the sources of some of the most wonderful (and most gruesome) stories of Catholic tradition, have been found by the Church to have been mythical figures.

Since 1234 it has been the Pope who proclaims a Saint. The performance of one or more miracles directly due to the intercession of a candidate must be proven prior to canonization. So the Vatican has established a tough multi-stage process, supervised by a Congregation for the Causes of Saints, working in Rome. The congregation has its own investigatory channels, but relies heavily on the opinion of a pool of medical experts (since nearly all miracles today are cures). At least two physician experts must aver that the ailment was not curable by any natural cause.

Until 1983, the curious figure of the Devil's Advocate (a defender of natural explanations!) played an essential role in the canonization proceedings. The process is now less adversarial but still arduous and slow; the Church has time. Many are nominated, few are canonized; a rational analysis of alternative natural causes of miracles plays a prominent role in the decision.[17]

#16
Date: 2 Feb
From: Roald Hoffmann (rh34@cornell.edu)
Subject: The prose of life

Your father-in-law, Shira, the late Yeshayahu Leibowitz, had an interesting point of view on the Marah
episode, and miracles in general. He noted first that miracles happen, and then pass, and "what endures is not the
exaltation of life, but rather the prose of life."[18]
Meaning the bitter waters, and the constant complaining of
the Israelites in the desert. He goes on to say:

> It is of tremendous importance for comprehending
> the essence of faith to grasp the fact that mira-
> cles and supernatural factors are of no signifi-
> cance religiously, and, at the least, are ineffec-
> tive as a foundation for faith. The generation
> which saw the miracles and wonders did not
> believe. . . . [In contrast,] there were many gen-
> erations in which the masses, not only unique
> individuals, clung to God and His Torah to the
> extent of laying down their lives for it. And
> those were generations which never experienced a
> Divine revelation, never saw miracles or wonders,
> had no prophets who spoke of God, and not only
> that, but God did not help them in their distress—
> and yet they believed.[19]

I think he has a point. And in another tradition,
Augustine, some fifteen hundred years earlier, remarked
that miracles were for fools who were more inclined to
rely on their senses than their intellect.[20]
But it may be that God speaks in several voices to
people, much as the head of the family at the Passover
Seder speaks to a wise son, and to an evil one, to one
who is simple, and one who does not know how to ask.
Those who have thought their way in to faith or are
secure in their belief do not need miracles. But perhaps
some of us do.

#17
Date: 7 Feb (Rosh Hodesh Adar, New Moon)
From: Kurt Jester (mercutio@verona.ivy.edu)
Subject: Miracles, bitter waters, and tenure

Sometimes our network discussions get too serious, shading over to lugubrious, for example, the bitter waters. Today is the first of the Hebrew month Adar, a day for jesting. Do you mind if I sweeten things up with a riddle? After creating the world and a few miracles, the Holy One blessed be He, decided to try for something really hard. He applied for tenure at a major Ivy League university. Tenure was denied, for three reasons. What were they? Answers in Bibl-e-mail next week.

#18
Date: 9 Feb
From: E. Z. Going (ezg@hal.oxbr.ac.uk)
Subject: Ramban's view

With respect to the complaint lodged by Mr. K. Vetch that we are digressing from our specific subject: The subject is the world, Vetch. And everything is connected to everything else.

Let me in fact introduce another discipline, linguistics, to shed light on bitter waters. In the context of Jewish religious study, careful attention to language was a hallmark of the work of the great Nahmanides, the twelfth-century religious scholar, physician, and philosopher. A case can be made for Nahmanides's anticipating the language theorists Ferdinand de Saussure and Charles S. Peirce.

In the old days couriers brought religious queries to the great rabbis from Jewish communities far away. I'm certain the Ramban would have taken well to e-mail. So here's a message from beyond:

> The meaning [of the passage] is that God showed Moses a tree and He told him, "Throw this tree into the waters, and they shall become sweet." Now due to the fact that I have not found the expression of *moreh* [from which the word *vayoreihu* here is derived], except in the sense of instruction . . .

> it appears by way of the plain meaning of Scripture
> that this tree had a natural property to sweeten
> water, this being its uniqueness, and He taught it
> to Moses.

After stating his own view, Nahmanides then cites the
explanation of the *midrashic* scholars and tries to show
how their double-miracle view is compatible with his "God
taught Moses."

> Our Rabbis have said that the tree was [naturally]
> bitter, but that this was a miracle within a mira-
> cle [i.e., that He healed the bitter waters with
> something which was bitter], just as the salt
> which Elisha cast into the waters. Now if so, the
> word "*vayoreihu*" (and He instructed him) indicates
> that the tree was not found in that place, and the
> Holy One, blessed be He, taught him where it was
> to be found, or perhaps He made it available to
> him by a miracle. I found further in the
> Yelamdeinu[21] "See what is written there: *Vayoreihu
> Hashem etz*. It does not say *vayar'eihu* (and He
> showed him) but *vayoreihu*, which means that He
> taught him His way." That is to say, He instructed
> him and taught him the way of the Holy One,
> blessed be He, i.e. that He sweetens the bitter
> with the bitter.[22]

I quote the Nahmanides passage in full not only because
it gives us the flavor of his reasoning, but also for the
way the personal enters his text—you can just feel
Nahmanides searching for meaning, can't you?

Also noteworthy is Nahmanides's plain reading of the
Bible—he simply accepts that the tree naturally sweetens
water. So it behooves us to find that tree.

#19
Date: 4 Adar I (11 Feb)
From: Shira Leibowitz Schmidt (shiras@netvision.net.il)
Subject: Chemistry lessons

I had recently been teaching English for scientists,
and had just gone with my class through an awfully complicated chemical article. I then thought I would take refuge
from all that chemistry, and escape to Jerusalem to a

class of Nehama's (oh, my aunt, Nehama Leibowitz, sadly no longer with us, was a wonderful Bible teacher and commentator) . . . only to walk in and find her talking about chemistry, too. She was giving a class on ibn Ezra's concept of miracles, and cited just these conflicting interpretations of the Marah episode that we have dealt with. When she came to the Nahmanides passage that E. Z. Going mentions, she summed it up: "So God gave Moses a chemistry lesson!" Or one in botany—where to find a rare species of tree. . . .

#20
Date: 13 Feb
From: Professor Yoram Zweifler
(yzweifler@immer.sicher.ac.il)
Subject: Nahmanides and a reply to the riddle

Nahmanides, though religious, is a man after my own heart.

I've done a little research on bitterness, and have located a book edited by Russell L. Rouseff—*Bitterness in Foods and Beverages*.[23] The book claims to be "the most comprehensive text devoted to bitterness in foods," but if I were to go by some of the misspellings and lack of documentation (one of the authors states that Marah was the Oasis of Raphidim, with nary a hint of the source of this discovery), the field is in a sorry state.

Still there are some interesting things in this book. In one article, J. A. Maga remarks that no general structural relationships have been established for bitter compounds due to the fact that the structures of known bitter compounds vary so widely. However, it is mentioned that bitter taste intensity correlates with the degree of hydrophobicity of a molecule and, in bitter-tasting aqueous solutions, there is an inverse relationship with surface tension. Professor Hoffmann, whom I understand has been having some funding problems, might consider investigating this problem theoretically.

Several chemical processes are described in this book for "debittering" food products. The methods used include the removal of bitterness from soy using solvents, precipitating agents, and microorganisms. Limonoid-metabolizing bacteria and physical adsorption are used to remove the bitter taste from citrus juices (but grapefruit juice had better be just a bit bitter!). None of this is relevant to

Marah, but it confirms my notion that it is possible to counteract bitterness by perfectly natural means.

May I switch to something else? I'd like to submit an answer to Kurt Jester's riddle about why the good Lord did not get tenure. It's obvious: He had no colleagues expert in his field to write evaluations of his work.

#21
Date: 16 Feb
From: Roald Hoffmann (rh34@cornell.edu)
Subject: Algistatic straw brews and ferruginous water

Of course, I like Nehama Leibowitz's idea of God teaching chemistry to Moses. I recently finished teaching about a thousand students in an introductory chemistry course, and with apologies to Nahmanides, I suspect God must have said "How nice to have a student who is not a pre-med!"

I just came across an interesting exchange on the waters of Marah in, of all places, the chemical literature. In *Chemistry in Britain* there was a little article about rotting straw stopping algal blooms. Apparently a farmer observed that the growth of algae in a lake had decreased the summer following a winter when some old and rotten bales of hay had fallen into the lake.

This observation encouraged Pip Barrett and his colleagues at the Aquatic Weed Research Unit in Sonning on Thames to study the matter "more scientifically." They found that as little as 10 grams of straw per cubic meter of water works, and are focusing on the hypothesis that the straw (barley is better than wheat) liquor is "algistatic," that is, contains some sort of plant growth regulator.[24, 25, 26]

This article elicited a response from S. Fletcher, C. Chem., F.R.S.C. in Fort Melbourne, Australia, who took the initial report to task for crediting "this kind of biological control to a certain un-named farmer, for in Exodus 15:23–25 we read [here follows our passage]. . . . It would appear that, in this case, Moses had priority!"[27]

That's not all. A couple of issues later, there is another letter in *Chemistry in Britain* from P. Day, C. Chem., M.R.S.C., from Widnes, Cheshire:

> It is more likely that the waters of Marah referred to in Exodus 15:23–25 were contaminated with

iron rather than algae as Dr. S. Fletcher
suggests. . . . Though I have never tasted water
contaminated with heavy algal growth, I can report
that ferruginous water is decidedly bitter!

I have always thought that the sweetening of the
waters of Marah resulted from the removal of the
iron by the coagulation and flocculation of
iron/tannin complexes, the tannin being derived
from that tree—though I think he [Moses] would have
had to have bashed it up a bit first.

Thus, we have chemical control—albeit naturally
derived—of a chemical contaminant, not a biological
one, and probably the first recorded instance.
Let's hear it for the farmer.[28]

Maybe the explanation is here! I might add that in the
last century tasting newly made or unknown chemicals was
recommended analytical practice. Not today, so I will not
ask Bibl-e-mail readers to tell me their experiences.

Perhaps the problem of Andy Goldfinger (he complained
in #4, 23 Jan, about everything tasting bitter) is that he
may have some steel-capped teeth (typical of Russian den-
tistry) in which the iron is deteriorating, really rust-
ing. And so he might be tasting, if P. Day is right, iron
ions.

#22
Date: 20 Feb
From: Norman Goldberg (ng20@cornell.edu)[29]
Subject: Another chemical suggestion

I am a postdoctoral associate in Roald's group at
Cornell, and he showed me your exchange on Bitter Waters.
I wonder if you've considered "Bittersalz?"

This is the old German name for magnesium sulfate,
$MgSO_4$. This chemical occurs naturally in the minerals
epsomite and langbeinite; perhaps one of these was at
Marah. Magnesium sulfate is quite soluble in water, 260 g
per liter in cold water. When it goes into solution it
gives magnesium and sulfate ions. Although I never tasted
it, I presume a solution of magnesium sulfate does taste
somewhat bitter.

Two other magnesium salts, the oxalate and carbonate, are much, much less soluble. What if the "tree" contained large concentrations of oxalic acid, or carbonate? There is oxalate in rhubarb leaves, for instance. Throwing the tree into the water might provide a source of these ions. They would combine with the magnesium ions, precipitating out of the water as solid salts, and presumably decreasing the bitterness of the lake.

#23
Date: 23 Feb
From: Geo. Witz (geowitz@cartog.co.il)
Subject: Location of Marah

Professor Yoram Zweifler (20 Jan) asks about the possible site of Marah. There are at least two theories about the route taken by the Israelites through Sinai, the southern and northern routes. [See illustration below.] For the southern route, Ain Hawarrah, a bitter lake on the eastern shore of the Suez Canal, is a likely candidate for Marah. Those who advocate the northern route to the Promised Land point to a body of water called Sabkhet el Bardawil.

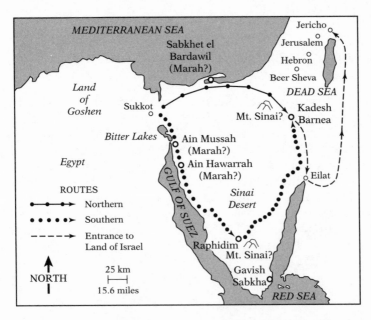

A map of the possible routes of the Israelites through Sinai.

#24
Date: 25 Feb
From: Fred Menger (menger@emory.edu)
Subject: Willows

I have an explanation for the "bitter waters" story. The water was basic, and hence bitter, as is often the case in the desert. Moses threw willow branches in the water, whereupon the alkali was neutralized by the salicylic acid in the willow bark. Willows, of course, are a likely plant to be found along streams. Salicylic acid is a precursor of aspirin, so Moses' headache was cured in more ways than one.

My theory, like all good theories, is amenable to experiment. All that is needed is alkaline water, willow branches, and a pH meter. However, I do not plan to do the experiment because I much prefer the Biblical version. God took pity on a forlorn group of people and gave them a helping hand.

Don't think ill of me, but I often find myself preferring the nontechnical explanation. I'd rather view the Northern Lights as did the Indians (i.e., as the campfires of dead ancestors) than as electrical storms on the sun.[30]

#25
Date: 28 Feb
From: Roald Hoffmann (rh34@cornell.edu)
Subject: Re: Menger suggestion

Fred: Glad to see still another chemist in this exchange! That's a great theory, friend. I have been told that bases (chemical, not baseball) are supposed to be bitter. We don't taste them the way they used to. . . . Are bases in fact bitter?

#26
Date: 3 Mar
From: Fred Menger (menger@emory.edu)
Subject: What we do for science

Basic water tends to be bitter. Toothpaste with carbonate in it has a bitter taste. I have just sampled a pH =
10 buffer (this is as high as I am prepared to go in the
interest of Science) and it seemed to be slightly bitter.

Here is another theory: Moses burned trees to charcoal
which he used to purify the water of bitter organics.
Charcoal is fantastic at extracting organics.

To solve your problem you need a collaboration between
botanists, physiologists, chemists, and geologists. I do
not recommend this, however. It is much more fun to operate, and argue, without facts.

#27
Date: 30 Adar I (9 Mar)
From: Ira T. B. Liever (ir8@2xs.com)
Subject: The location of Marah

There is a third school of thought, which identifies
Kadesh Barnea as the site of Marah. Now Mr. Witz, why
don't you cite the source of your water source? I found it
listed in Encyclopedia of the Bible,[31] and even mentioned
in a midrash,[32] which asks not "which tree?" but "where?"

And, incidentally, how do you know that Ain Hawarrah is
bitter?

#28
Date: 11 Mar
From: Geo. Witz (geowitz@cartog.co.il)
Subject: Lakes and silly question by Liever

Because I tasted it, you dummy, when my army unit was
stationed near that inhospitable waterhole in 1970.

#29
Date: 12 Mar
From: I. M. Shammes (News group moderator)
(shammes@yu1.yu.edu)
Subject: Proper behavior in Bibl-e-mail

    OK, gang. Let's restore some civility in our dialogue.
Or you'll be banished to e-mail Sodom and Gomorrah.

#30
Date: 4 Adar II (12 Mar)
From: Ira T. B. Liever (ir8@2xs.com)
Subject: Bitters

    I'll stop, I'll stop. But if Witz is so smart, he
should have analyzed the waters of the lake when he was
last in Sinai, so we'd know what made it bitter. . . .
    I still believe that it doesn't matter just what made
Marah bitter, or what Moses cast in the water. And whether
G-d caused the miracle or taught Moses. Any means would
have sufficed.
    I do respect the Ramban's search for the word of G-d.
The sages were quick to utilize linguistic ambiguities and
connections to extract ethical import. *Mar* is the Hebrew
root for bitter, as in Marah, and also as in *marror*, the
bitter herbs eaten at Passover. The Bible says, ". . . they
could not drink the waters of Marah because they were
bitter. . . . Some sages comment: "because *they* were
bitter . . . the people of that generation."[33] They certainly did complain a lot, beginning at Marah. They should
have put their faith in G-d.

#31
Date: 12 Mar
From: Wolfgang E. Krumbein
(wek@africa.geomic.uni-oldenburg.de)
Subject: Sabkhas

    I think Norman Goldberg and Geo. Witz are on the right
(bitter) track. I've worked on *sabkhas,* and I bet that it
is one of these that the Israelites encountered. A *sabkha*
is a salty environment, a dry lake, lagoon, or depression

that periodically gets wetted or flooded. *Sabkhas* may be covered with treacherous white salt efflorescences, or they may have brown-colored spots on the surface, with a slimy layer of cyanobacteria under the salt.

One of these I've worked on is the Gavish *sabkha* (named after an Israeli sedimentologist who died prematurely in 1981).[34] It's on the shore of the Red Sea, near the bottom of the Sinai. The center of the Gavish *sabkha* is elevated and the water is pumped in by the sun and pumped up by the sun. So the center is not a depression but an elevation (compare it to electron holes and electron clusters). To reach the center you must pass gypsum and mud (stinking). After it your legs are scraped parallelly red from ankles to above the knees by the gypsum layers which do not hold your weight. The center tastes bitter!

This *sabkha* is not where the ancient Israelites went, but there are other such "bitterns" in the area. These are salt pans with bitter-tasting waters in their central parts. Bitter because the sodium chloride has crystallized out, or washed out, or has been collected by humans (my expert Bedouin told me the bitter taste occurs when the salt is harvested). The bitter magnesium salts are then concentrated in the middle.

Although Roald Hoffmann's explanations are more elegant, Norman Goldberg is my favorite man for straightforward hypothesis.

Now for the cure! The Moringa tree (*Moringa peregrina*) is widely distributed in Africa and around the Red Sea. It was (in biblical times) used for fences, as a pleasure plant, a vegetable, and for many other purposes, among which is . . . cleaning water.

The tradition of cleaning the water of the Nile with the seeds of the Moringa tree was already used by the Nubians. Via Imhotep the technique came down to scientists of the European community, who recently have run a project in Malawi—they use the Moringa seeds first for oil production, and then sell the oil cake, which still contains proteins serving as sorbents for bacteria, clay, and salts, for water purification. This technique has been widely used among the people living along the Nile in order to clean the water, which might be polluted, dirty, or of bitter taste. No wonder that Moses told the "people" to throw a tree into the bitter waters![35]

#32
Date: 14 Mar
From: Norman Goldberg (ng20@cornell.edu)
Subject: More support for Bittersalz

I just found a fascinating book by Arie Issar, *Water Shall Flow from the Rock*. Issar states that the large freshwater springs of Ain Mussah (Uyun Musa, "Springs of Moses" in Arabic) [see the illustration below], not far from proposed Exodus routes, in fact contain magnesium sulfate, "which gives a bitter taste."[36]

#33
Date: 15 Mar
From: E. Z. Going (ezg@hal.oxbr.ac.uk)
Subject: Mr. Liever's comments of Mar 12

Mr. Liever's point was that the people were bitter. This moral reproach is possible because of the confusing

One of the springs of Ain Mussah (Uyun Musa). (Photograph by Emmanuel Mazor.)

syntactical structure of most languages. So Nehama Leibowitz compares the Marah verse to another ambiguous phrase in English—juxtapose these two sentences:

(a) The Isrealites   could not drink the waters of Marah   because **they** were bitter

(b) The professors   could not teach the students   because **they** were stupid

In trying to understand (a) we ask: the Israelites, or the waters? In (b) we ask: the professors, or the students?[37]

The difference between simply ambiguous syntax and the biblical example at hand is that the alternative meaning (the people being bitter) is one with psychological depth and significance.

There are undercurrents in the *midrashim*, interesting to follow, and leading us to other passages in the Bible. Among the sages of the "bitter sweetens bitter" school, one for instance posited that the olive tree provided the branch. He assumes the reader may remember the olive branch from the Noah's Ark episode. There the dove, according to one *midrash*, implied: "better my food be bitter as an olive leaf from the bitterest of trees, but be eaten in freedom, rather than be pleasant tasting but eaten in the cage of the ark."[38] Then you have a different tradition that sees an abstract metaphorical connotation to the tree: The tree cast into the lake was the Tree of Life. That tree is the Torah and its teachings, which were, and are, the antidote to the bitterness.[39]

#34
Date: Purim (23 Mar)
From: Kurt Jester (mercutio@verona.ivy.edu)
Subject: The answer you've been waiting for

Professor Yoram Zweifler made a nice try at answering the Creation miracle riddle. And we should give him an electronic hand for lightening up!

I had in mind three reasons for the rejection of God's application for tenure:

1. No one has ever been able to repeat His experiment.

2. God has only one publication, and that was in a nonacademic, unrefereed venue.

3. His first two experimental subjects were expelled.

#35
Date: 24 Mar
From: Roald Hoffmann (rh34@cornell.edu)
Subject: Taste modifiers

Hey, Bibl-e-mail gang, I just got a great idea. Did you know that there are substances that are "taste modifiers"? They have little or no taste themselves, but they modify the taste of what is eaten or tasted after the tongue is exposed to them.

First a few words about taste. We experience this sense through "taste buds" located mainly on the tongue, but also found on the soft palate and epiglottis. In the vicinity of these buds, the thick overlayer of the tongue (keratin) becomes very thin, and an opening, a pore, is formed. Through that pore, liquids and small particles can approach the microvilli which are part of the receptor cells that may eventually fire a nerve signal. [See the illustration on page 148.]

There are receptor cells sensitive to different tastes, so it is not easy to make sense of taste. Human beings are also prone to mix taste and smell, especially in their perception and description of foods.

Be that as it may, certain taste sensations are pretty identifiable; they may be thought of as the coordinate axes of a multi-dimensional space in which the complex taste of a pomegranate or a bagel might be described. The clearest such taste axes are sweet, salty, bitter, and sour.[40,41]

There is more, and I can't say it any better than Diane Ackerman in her wonderful book, *A Natural History of the Senses*:

> The tongue is like a kingdom divided into principalities according to sensory talent. . . . A flavor traveling through this kingdom is not recognized in the same way in any two places. If we lick an ice cream cone, a lollipop, or a cake-batter-covered finger, we touch the food with the tip of

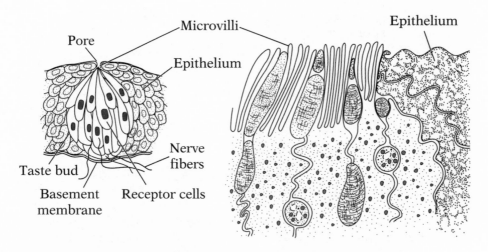

A schematic of the human taste apparatus.

the tongue, where the taste buds for sweetness are, and it gives us an extra jolt of pleasure. A cube of sugar placed under the tongue won't taste as sweet as one placed *on* the tongue. Our threshold for bitter is the lowest. Because the taste buds for bitter lie at the back of the tongue; as a final defense against danger they can make us gag to keep a substance from sliding down the throat. Some people do, in fact, gag when they take quinine, or drink coffee for the first time, or try olives. Our taste buds can detect sweetness in something even if only one part in two hundred is sweet. Butterflies and blowflies, which have most of their taste organs on their front feet, need only step in a sweet solution to taste it. Dogs, horses, and many other animals have a sweet tooth, as we do. We can detect saltiness in one part in 400, sourness in one part in 130,000, but bitterness in as little as one part in 2,000,000. Nor is it necessary for us to recognize poisonous things as tasting different from one another; they just taste bitter.[42]

On to taste modifiers. *Gymnema sylvestre* is a tropical woody climber. Its leaves are used in traditional medicine in India, but its most spectacular property is evident in its common name, *gur-ma,* or "sugar-destroying."[43] *Zizyphus jujuba* is a deciduous small tree or bush of warm regions, whose leaves have a similar sweetness modifying effect.[44] For some minutes after the relatively tasteless extracts of these plants are chewed, the perceived sweetness of sugars nearly vanishes: ". . . an orange eaten after one has chewed Gymnema leaves tastes like a lemon or lime."[45]

The chemical structures of the active principles of these plants are known. [One of them, ziziphin,[46] is shown below.] God didn't pass up a chance to teach chemistry; I won't either.

The moderately complicated $C_{51}H_{80}O_{18}$ molecule is of a class called glycosides. The "glyco" refers to a sugar piece (but only some of the class of molecules chemists call sugars taste sweet). In ziziphin there are three "sugar" rings—those six-sided rings with lots of oxygen. The remainder of the glycoside is a nonsugar piece, either a steroid, or a class of molecules called triterpenes. That's the set of carbon hexagons sharing edges in the center of the molecule. The glycosides in the antisweet principles are called saponins, for they foam when shaken

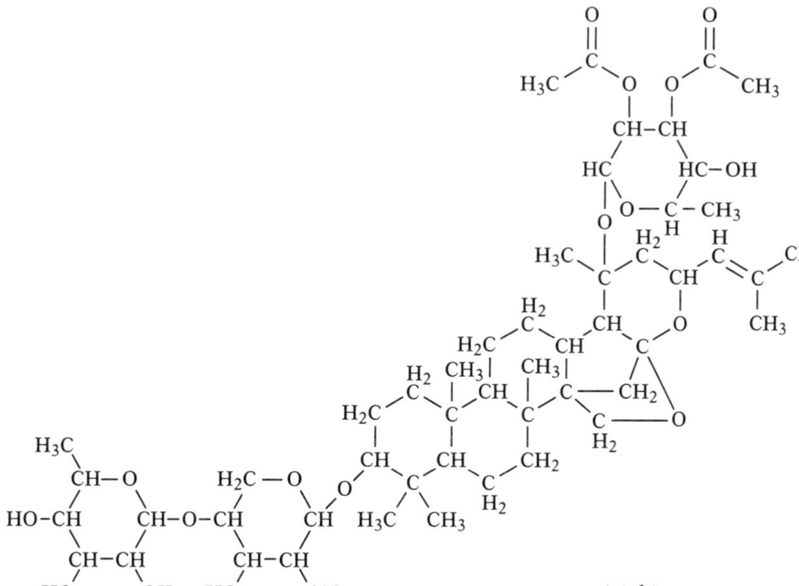

The chemical structure of ziziphin, a sweet-taste modifier.

in water and form emulsions with oils, like soaps.
Saponins have a lot of biological activity—some are toxic
to snails, some show antitumor activity. Some are sweet,
some bitter, some have been commercially important as precursors of steroid hormones.[47]

There is another sweet taste modifier that is found in
the berries of an African shrub, *Synsepalum dulsificum,*
and called *miraculin,* or the miracle fruit. It has the
opposite effect—to a person who has chewed a *Synsepalum*
berry, a lemon tastes like an orange. For a few minutes.
In the literature, similar effects have been reported for
artichokes.[48]

The great majority of well-established taste modifiers
affect the sweet taste; only recently have compounds
inhibiting bitterness been identified.[49] Here's my idea:
On the shores of Lake Marah there grew a species of tree
that was a bitter taste modifier, the better debitterer.
It inhibited for the thirsty Israelites the bitterness, of
the really safe-to-drink water, wherever that bitterness
came from. What do you think of that?

#36
Date: 15 Adar II (24 Mar)
From: Ira T. B. Liever (ir8@2xs.com)
Subject: *Too much* chemistry

   Another disciple of the great professor Yoram Zweifler!

#37
Date: 31 Mar
From: E. Z. Going (ezg@hal.oxbr.ac.uk)
Subject: Taste modifiers

         Ah,
         Those who call evil good
         And good evil;
         Who present darkness as light
         And light as darkness;
         Who present bitter as sweet
         and sweet as bitter

                    Isaiah 5:20

#38
Date: 31 Mar
From: Perennial K. Vetch (peryk@sr3.catnet.org)
Subject: Too much

I know, *rabossai*, that we all like to talk. But Bibl-e-mail comments that go on, and on, . . . My computer has a tiny monitor, and all these grandstanding speeches use up so many lines on the screen that I lose track of points made by my other learned friends. Let's adopt a rule limiting a person to twenty lines of text. And no chemical formulas. Or let's organize a separate news group for biblical chemistry and get these guys out of our hair and off the air.

Stick to the most *salient* points about bitter water!

#39
Date: April Fool's Day
From: Code Bleu® (www.codebleu.com/boost)
Subject: Just what you need

Fellow seekers of miracles! We invite you to visit our World Wide Web Site, to try on a pair of Miracle Boost™ jeans in our "virtual fitting room."

Trying on Miracle Boost jeans is easier than going to your local department store. No lines, no parking hassles, no stepping on pins. All you need to do is

1. enter your size—don't worry: they scroll down;

2. select the color you want—dark indigo, medium stonewashed indigo, black;

3. push a button.

#40
Date: 1 Nissan (8 Apr)
From: Shira Leibowitz Schmidt (shiras@netvision.net.il)
Subject: Other bitter waters

In trying to clear up the murky problem of bitter waters of Marah we have waded deeper and deeper into that passage. I suggest we explore other biblical passages that may add dimensions to our quest.

The second biblical mention of bitter waters is in a very different, psychologically rich context, the trial by

ordeal of the suspected adulteress, called *sotah*, in
Numbers Chapter 5. Pathologically jealous husbands are not
an invention of Restoration comedy, but a feature of every
society. And they are not funny at all—such jealousy,
often without cause, can ruin a family. Jewish tradition
provided what I view as a perceptive rite to heal such
jealousy:

> The priest shall take sacral water in an earthen
> vessel and, taking some of the earth that is on the
> floor of the Tabernacle, the priest shall put it
> into the water. . . . And in the priest's hands
> shall be the water of bitterness that induces the
> spell. The priest shall adjure the woman, saying to
> her "If no man has lain with you, if you have not
> gone astray in defilement while married to your
> husband, be immune to harm from this water of bit-
> terness that induces the spell. But if you have
> gone astray while married to your husband . . . may
> the Lord make you a curse and an imprecation among
> your people, as the Lord causes your thigh to sag
> and your belly to distend.". . . The priest shall
> put these curses down in writing and rub it off
> into the water of bitterness. He is to make the
> woman drink the water of bitterness. . . .[50]

[A seventeenth-century image of the beginning of the
sotah ceremony is shown on page 153.[51]]

The *Zohar* (the basic *kabbalistic* text) actually makes a
direct connection between the bitter waters of Marah and
those of the suspected adulteress. It quotes R. Eleazar,
who comments on the biblical line right after Marah:
"There He made for them a fixed rule, and there He put
them to the test." Eleazar says the test was an ordeal of
bitter waters for both Israelite men as well as women, for
both sexes were accused of consorting intimately with
Egyptians. So the bitter waters of Marah here are the ritual ordeal waters for a whole nation.[52]

There is another passage in the Bible in which a transformation of unpotable water is effected. It is accomplished not by a tree but by the prophet Elisha's adding
salt (!) to the water—another miracle within a miracle
(II Kings 2:21). As in our case at Marah, we are not told
what fouled the water. At Jericho the inhabitants simply

The Ordeal of Bitter Waters. Engraving by C. N. Schurtz, 1674.

state that the water is "bad" *(raim)*. Elisha then "cast
the salt in and said: Thus said the Lord: I have healed
these waters."

Is there a moral connotation to "bad" waters that "bitter" doesn't have? And note that the concept of healing
(rifeyti) is brought to bear in the Elisha story, just as
it appears in the conclusion of our Marah episode, where
God says "I the Lord am your healer." [53]

#41
Date: 15 Apr
From: Moshe Greenberg (greenberg@mail.cjs.upenn.edu)[54]
Subject: The incident at Marah

I can report only the view of a nonscientist. The story leaves open whether or not the tree/wood healed the water because of a natural property. Ramban makes a fine point in pointing out that *vayoreihu* suggests some sort of instruction about the natural healing property of the substance (which Moses wouldn't have known otherwise).

Comparison with Elisha's healing the waters of Jericho shows that Moses' knowledge of the healing substance came from God (salvation is of and reflects glory on God) while Elisha's was occult knowledge he had on his own (salvation is of the man of God and reflects glory on him). Again there is no direct answer to the question whether Elisha knew something about the natural property of salt that nobody else knew or whether the crucial healing element was supernatural and ad hoc.

The aim of both stories is to glorify God (Marah) or "his man" (Jericho) and that is accomplished by ascribing to each the ability to heal water in a way nobody knew (before). The question at issue could be settled if afterward an ordinary person could achieve the same result. But for that we should have to know exactly what embittered the water, exactly what tainted the Jericho water, and exactly what the composition of Elisha's salt was. Once we determined these variables we might pronounce on the theoretical possibility of the events narrated, but that would not change the supernatural origin of Moses' knowledge of the property of the tree/wood (specified in the story) or the extreme improbability of Elisha's having knowledge of the exact substances and proportions that enabled him to purify Jericho's waters—an improbability that requires the assumption of a supernatural source of such knowledge to make it plausible.

So the pronouncement of some chemist-juggler on the theoretical conditions for accomplishing water-purification as described in these stories cannot affect the primary/ultimate postulate of these stories that a supernatural event is narrated in them. To think that by showing the theoretical possibility of the events having occurred naturally one has enhanced the credibility of the Bible is vain, since what the Bible asserts is not merely that such events happened, but that they attest the providence of a supernatural God at work on behalf of his people directly (God instructs Moses) or indirectly (through "his man"). The more natural one makes these events, the more one undermines the Biblical depiction of them as evidence of God's care of his people.

That is why this nonscientist bridles at attempts by scientists to rationalize or naturify Biblical miracles. Such attempts, while undertaken in a spirit of piety, defeat the very purpose of the Bible in reporting them. What place a faith in miracles has in the belief of moderns is a separate serious issue that no eviscerating rationalization of Biblical miracle stories can help to determine.

#42
Date: 18 Apr
From: Roald Hoffmann (rh34@cornell.edu)
Subject: Miracles, and a pox on both your houses

I like what George Bernard Shaw said, and in his spirit, may I make a few comments irreverent of both science and religion?

I think religion and science are both attempts by human beings to come to terms with the beautiful but quirky universe. In any era, in any day of our lives, we are faced with much that is clear and a range of happenings that we do not understand—the wisest and kindest of women is killed by a hit-and-run driver, a tumor that should kill recedes. If one understands (or desires) the event, one calls it explainable by science, or the answer to a prayer. Even a miracle. If one doesn't understand, then the scientist renames it as a "spontaneous remission" of the tumor—the demon named is tamed. And if one is religious, and if the event is blatantly against all values (as in the death of that woman, or so many of my people in World War II) then one falls back on the book of Job, and the unfathomable will of God. But if the event is desirable, it's a miracle. How self-centered that last view is, how not of God, but of Man!

I say a pox on both your houses. No, I really don't mean that, because I love you both, as much as I love the unfathomable path of the tornado. I will take the beauties that scientific understanding brings, its infinite

intricacies and simple equations alike. And I will accept
the sublime, and the ethical core of religion. Both of you
are *so* deep. And you are *so* obvious when you try to
stretch the wonderfully hazardous web of what *is* into the
flimsy simple cloth of what humans want to be.

#43
Date: Passover Eve (21 Apr)
From: Shira Leibowitz Schmidt (shiras@netvision.net.il)
Subject: The text around

   I think our discussion has run its course.[55] Marah, or
any miracle, is guaranteed to elicit the immediate
responses that I got to my query: "There must be a scientific explanation!" and "Who cares, it was a miracle!" As
we saw, the ground in between filled in—we had ancient
and modern Jewish religious thinkers very much concerned
about the precise nature of the tree. And we have seen
scientists confronting the everyday miracle of taste,
sweet or bitter. Perhaps in the end the opponents in this
debate have not been shaken from their initial convictions.
But I think that they, and I know I, have learned a lot.
   It is always useful to look at a biblical passage in
context. The verse I cited at the beginning comes right
after the passage of the Red Sea, a high point at which
Miriam, Aaron's sister, and the women of Israel dance
and sing.
   But life goes on, never stopping long to savor celebration. As it is in the world, so it is in the Torah. The
people have a long way to go to the Promised Land, it's
time to move on. Then comes Marah, the very first of a
long sequence of tribulations. Each of these will lead to
dissatisfaction if not rebellion, followed by a miracle,
and conclude with an admonition to hearken to His ways.
For Marah, here are the verses right after the episode of
the tree that Moses threw into the bitter waters:

> There He made for them a fixed rule, and there He
> put them to the test. He said, "If you will heed
> the Lord your God diligently, doing what is upright
> in His sight, giving ear to His commandments and
> keeping all His laws, then I will not bring upon
> you any of the diseases that I brought upon the
> Egyptians, for I the Lord am your healer."[56]

The Lord our healer lays out the compact pretty straight. But we the people of Israel are . . . just people. We will not learn.

Perhaps it's a drop better than this. Why do we need Moses, or the tree, at all? Obviously, without denying the omnipotence of God, there is in the Bible a general principle at work of enlisting human cooperation in the work of the Deity. This idea is found in many places,[57] and it is perhaps the reason why Nahmanides felt the key phrase in the Marah passage was "and the Lord instructed/showed him. . . . " We are partners, albeit unreliable ones, in a sacred compact.

It is Passover eve as I type this e-mail message, time to remember the Exodus.

The Flag That Came out of the Blue

A Play in Three Acts and Two Intermezzi

ACT I*

SCENE 1. *It is dawn. Against a backdrop of low desert mountains, we just begin to discern hues of brown and scattered green. Onstage are three tents, more merge into the barren landscape. Out of the rightmost tent, the simplest one, Moses emerges. In his hands he bears a four-cornered white poncho-like garment, with holes in its corners through which he is painstakingly threading strings of white and blue. Israeli Scouts today can use his technique to earn their merit badge in knot-tying. [See the illustration on page 160 for a related contemporary scene.]*

Moses finishes, dons the garment and gathers the fringes in his left hand. He looks questioningly at the sky, which is growing lighter by the minute. He looks back at the fringes, to see whether he can tell the difference between blue and white, one of the traditional talmudic tests for sunrise.[1] It is time. He then recites this passage from Numbers 15, which has been incorporated into the morning and evening prayer service.

Moses: And God said to me, He said: "Speak to the Israelite people and instruct them to make for themselves fringes (*tzitzit*) [*kisses*

---

* Nearly every word spoken by the characters in Act I comes from Numbers 15, 16, or 17, or from Midrash Numbers Rabbah. The original texts lack stage directions, so these are ours.

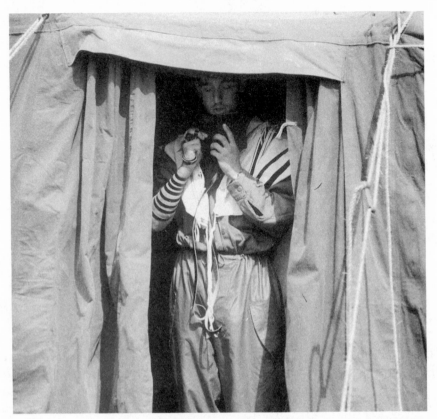

Not Moses, but a contemporary Israeli soldier praying at dawn by the door of his tent. Note the prayer shawl, *tallit,* over his shoulders and the fringes (all white) at its corners. The leather straps of his *tefillin* (phylacteries) are wound on his arm and hang down.

them] on the corners of their garments throughout the ages; let them attach a cord of blue (*tekhelet*) to the fringe at each corner." [*He turns to the group of ten men and boys who have gathered around him, and holding up the fringed garment, he continues*]: "That [*Moses raises up high the bunched fringes*] shall be your fringe; look at it [*he passes them before his eyes*] and recall all the commandments of the Lord and observe them, so that you do not follow your heart and eyes in your lustful urge. Thus you shall be reminded to observe all My commandments and to be holy to your God. I the Lord am your God, who brought you out of the land of Egypt to be your God: I, the Lord your God." [*He kisses the fringes.*]

[*While Moses speaks, from the grandest of the three tents, the multicolored one at left, Korah, a greatgrandson of Levi, emerges, along with his wife. They are quiet as they watch Moses pray, but there is tension in their bodies. Korah was controller in Pharaoh's house, he held the keys to Pharaoh's treasury. Korah is a clever man. This is why his tent stands out above the others.*]

Korah's Wife: [*to her husband, quietly*]: I remember so clearly the day he said "Take the Levites from the Israelites and cleanse them. This is what you shall do to them to cleanse them: sprinkle on them water of purification, and let them go over their whole body with a razor, and wash their clothes; thus they shall be cleansed."[2]

Korah: I will not forget that day; he did it to me himself, with his own hands. He grabbed me by my hands and feet and I was swung in the air, naked, and he told me: "Behold, you are clean!" This he did to me, in sight of all the others. Afterward I walked among the Israelites and . . . they didn't recognize me. My friends said to me: "Who did this to you?" I answered: "Moses did it to me."

Korah's Wife: As if this were not bad enough, he cut your hair off— he is making sport of you, because he despises you; for some reason he envies your hair. Then he brought his brother Aaron and decked him out like a bride and made him sit in the Tent of Meeting!

Korah: Moses is king, his brother Aaron is high priest, and his nephews are deputy high priests! The priests get the heave offering, the priests collect the tithe, the priests get twenty-four gifts!

Korah's Wife: How long shall this go on? Korah, you are a man of great subtlety, of wealth, you can *do* something about this.

[*Korah thinks, looks at Moses caressing the blue fringe. Korah looks at the sky, now all blue. He leaps up.*]

Korah: I *will* challenge him, with just this blue fringe that he has put on us. Listen to me, woman—gather my other wives, and your daughters and servants. Make for me 250 cloaks, all of *tekhelet* blue throughout, and I will seek out my friends, 250 men of repute, chieftains of the community, all chosen in the assembly. They despise Moses and his pretensions! We will wear the cloaks as one and rise against him.

[*Wives, daughters, and servants surge in. In a tumultuous scene, carts bring in live snails in buckets; the snails are broken apart, a liquid extracted, treated with some chemicals. Elsewhere on stage, wool is brought up, wool sheared from the many sheep in Korah's flocks. The wool is dyed right on stage in deep vats—a maid holds up a skein to the sun, and it turns purple before our eyes. The wool is spun, woven. The stage is all blue and purple, the sheep overturn the vats, the skin of the people glistens with the dye. At center stage Korah's wife sits down to her Singer sewing machine, with several daughters. She sings as she begins to sew the first of 250 purple cloaks.*]

SCENE 2. *It is night, and the stars shine on the desert. Korah moves stealthily from tent to tent.*

Korah: I say, no more special privilege, brothers. Do you suppose, that I do this to obtain greatness for myself? I desire that we should all enjoy greatness in turn, not like Moses who has appropriated the kingship to himself and has given the High Priesthood to his brother! All the community are holy. *All* the community are holy!

[*One of the eager listeners is Onn, the son of Peleth. He is a simple man, he hears Korah and is swayed to join the rebellion. After Korah leaves, Onn turns to his wife.*]

Onn: It is time that the tribe of Levi received its due, don't you think?

Onn's Wife: Husband, what will you get out of it? Either Moses remains master and you a follower, or Korah becomes master and you—you are still no more than a follower.

Onn: But what can I do? I have taken part in the deliberations, and I have sworn loyalty to Korah. . . .

Onn's Wife: Don't worry, Onn. Here, taste some of this old wine. . . .

[*She plies him with Chardonnay, vintage 1400 b.c.e., poured in a new flask, of course, puts him to bed within the tent, and seats herself at the entrance. She then loosens her beautiful chestnut hair, and lets it down. Dawn is breaking. One of Korah's troop comes to summon Onn and sees his wife with her hair down. He looks immediately away, for it is*] *immodest to look upon another man's wife with her hair loosened. Onn's wife sways gently, humming. Korah's follower hesitates, then runs off. Onn sleeps on peacefully while the rebellion takes place. A voice offstage sings: "'The wisest of women builds her house'—may be applied to the wife of Onn, and 'but folly tears it down with its own hands'—the wife of Korah."[3]]*

SCENE 3. *Inside Korah's big tent. Drums and trumpets offstage left. The conspirators enter. The smell of roasting meat fills the air.*

Korah: Friends, Israelites I lend you these cloaks—they will make all of us holy. But first do me and my family the honor of joining me in this feast. [*Turns startled to the door*] Sons of Aaron, what are you doing here?

Son of Aaron 1: We come to receive our dues . . .

Son of Aaron 2: the breast and shoulder of the animals that have been slaughtered for the feast . . .

Son of Aaron 1: as the Lord said to Moses in Leviticus 7:31 . . .

Son of Aaron 2: and the next verse, too.

Aviram, a Follower of Korah: Who told you that you are entitled to take these? Was it not Moses? We shall not give you anything! He who is everywhere, He did not command us to do such a thing!

Son of Aaron 1: [*to his brothers*]: They deny us our portion. Let us seek Moses' counsel.

[*He takes out a cellular phone (there is a higher density of cellular phones in Israel than anywhere else in the world), dials Moses, and relates what has happened. Moses walks over to the big tent. As he contemplates the rebellious crowd his face falls.*]

Moses: What is it, my people? Why are you dressed in this way?

Korah: [*who was seated, jumps up*]: Moses, if a cloak is entirely blue, what is the law? Should it be exempt from the obligation of a blue fringe in each of the four corners?

Moses: The garment still requires the fringes. So the Lord has spoken to us, through me.

Korah: Is that logical? A robe of any color fulfills the fringe requirement merely by having one of its threads blue. Surely a garment that is *entirely* blue should not require an additional blue thread! [*He looks around at his followers, who acclaim his logic.*]

These things you have told us—you have not been commanded concerning them, you are just inventing them out of your own mind. All the community are holy!

Moses: You have gone too far, sons of Levi! Come morning, the Lord will make known who is His and who is holy, and will grant him access to Himself; He will grant access to the one He has chosen. Do this: You, Korah and all your band, take fire pans, and tomorrow put fire in them and lay incense on them before the Lord. Then the man whom the Lord chooses, he shall be the holy one.

[*Moses leaves, but as he goes out of the tent and looks up at the sky which begins to color with the blue of morning, he says quietly*]:

Perhaps they have been speaking as they did because they have just eaten and drunk. Perhaps, between now and then they will repent. . . .

SCENE 4. *It is the morning after; the people gather solemnly to see the test of the 250 fire pans. Aaron and Moses are there too, facing the blue-clad rebels at the entrance to the Tent of Meeting. Suddenly a whirlwind comes, a thunderclap, radiating rings of fire. Indiana Jones is nowhere in sight. A voice speaks, which only Moses and Aaron hear.*

The Lord: Stand back from this community that I may annihilate them in an instant! [*Moses and Aaron fall on their faces.*]

Moses and Aaron: O God, Source of the breath of all flesh! When one man sins, will You be wrathful with the whole community?

The Lord: Speak then to the community and say: "Withdraw from about the abodes of Korah, Dathan, and Aviram."

[*Moses and Aaron turn to the people, and speak with unquestioned authority. Pandemonium, the people flee, leaving the families of the three rebellious clan heads awaiting judgment in front of their tents.*]

Moses: [*strong in the beginning and the end, a little weak in the middle*]: Lord, You are merciful. But hear me, once more. Korah said I have invented Your rule in my own mind. If these people die as all men do, if their lot be the common fate of all mankind, it will seem it was not the Lord who sent me. So they will say. But if the Lord brings about something unheard of, so that the ground opens its mouth wide and swallows them up with all that belongs to them, and they go down alive into Sheol, all shall know that these men have spurned the Lord.

[*The earth indeed opens up, swallows Korah and his fellow rebels. And a fire, not of Steven Spielberg's ken, consumes the 250 men in blue.[4] The curtain closes on Act I.*]

INTERMEZZO 1. *The audience wanders out from the theater, moving quietly to the foyer, where vendors sell Moroccan cigars, small sweetmeats. People are still stunned by the drama of Korah's rebellion; their talk is subdued. Not that they get much time for talk, for over a loudspeaker system there comes a clap of thunder, an echo of what they just heard in the theater. A small bald man emerges on a platform. He wears a dapper tropical white suit and a blue tie, and begins to speak.*

Professor Azul: Ladies and gentlemen; allow me to introduce myself. I am Professor Azul, from the Half-Shut University. Noting that theater intermissions are rather long, and that much time is wasted in idle talk, we have decided to offer our theatergoers the option of intellectual refreshment in the intermission. It is my pleasure to tell you a little bit of the background of this play, both scientific and religious. Should you prefer a glass of champagne, I will not at all be offended if you step outside on the lawn.

[*This being such a radical intrusion into their theatergoing experience, few in the crowd move to depart. The professor continues, with some satisfaction. Slides flash on a screen behind him.*]

Blue pigments were common in antiquity. My first slide shows an incised wall painting of the goddess Isis, with cow's horns and sun disc, from the Valley of the Kings, Luxor, dated to 1332–1305 B.C.E. Alexandria was the source of the best Egyptian blue, a powdered glass frit made from calcium copper sulphate. It was used throughout antiquity as a high-quality but less expensive substitute for true ultramarine blue, the blue from "across the sea" which had to be obtained laboriously by grinding the semiprecious stone lapis lazuli.[5] Most of the pigments used for decorating the walls of tombs and painting mummies' coffins and statues were of mineral origin. Metal oxides were also used to obtain the characteristic coloring of Greek and Roman wall painting.

When we come to textiles, the mineral pigments, by and large, fail to serve. Dyes of animal and plant origin were found for coloring textiles around the world, often using uncommon plants or animals. Many protochemical experiments probed the often elusive fastness of these dyes. The play you are enjoying features an unseen actor, one such dye, covering the range around blue.

From early on in the history of Rome (and as it will turn out, in the lives of the Hebrews) a purple-dyed wool, of a hue ranging from red to blue-black, was highly valued. It was called Tyrian purple or Royal purple. Pliny the Elder describes it as "the color of congealed blood, blackish at first glance, but gleaming when held up to the light."[6] In Republican Rome, clothing completely dyed in purple could be worn only by the censors and triumphant generals, while the consuls and praetors wore purple-edged togas, and generals in the field a purple cloak.[7]

The manufacture of Royal or Tyrian purple was highly restricted in the Roman Empire. In fact, by an edict of the emperors Valentian, Theodosius, and Arcadius in the fourth century, it became a capital offense to manufacture the Royal purple other than in the Imperial Dye-Works. The privilege of wearing the true purple (more to come on what's "true" and what isn't) was restricted to the emperor. In 301 C.E. a pound of Tyrian purple-dyed wool cost 50,000 denarii, about a thousand days' wages of a baker.

I note, parenthetically, that one thread of Islamic tradition explains the wealth of Korah through his wife, said to be Moses' sister. Moses taught her alchemy, and from her Korah learned to make gold. So it is said.[8] Tyrian purple was, weight for weight, more precious than gold. . . .

Long before Rome the Israelites hearkened to a prescription for a blue in the Bible. You heard it in Moses' own words—God's commandment to wear *tzitzit* (fringes), with a blue strand in the fringe. Consistent with the specificity of biblical and talmudic concepts of holiness, these blue strands had to be colored with *tekhelet*, which is a dye from a snail.[9] Just how important this prescription and its observance were to the Israelites in everyday life may be judged from the fact that the very first tractate of the Talmud inquires: "From which moment may one recite the Shema [the prayer of creed] in the morning?" And as the first answer gives: "When one can distinguish between *tekhelet* and white." This is what Moses was doing before he began speaking at the start of Act I.

In Jewish tradition, the lore of making *tekhelet* was lost by 760 C.E. Since then the prayer shawls of the orthodox have included no colored strands in their tassels, for there is no substitution in the eyes of the Law.[10] The Byzantine Imperial Purple Works, and the art of making Tyrian purple, vanished with the fall of Constantinople in 1453.

Meanwhile the world turned, many times. Blue and purple were or became available from other sources, not the least of which, from the end of the last century, is that piece of human ingenuity called chemistry. And we have also found out what natural Tyrian purple and biblical blue were.[11]

These pigments are of animal origin. They were extracted painstakingly, and therefore expensively, from three species of gastropod snails: *Murex brandaris, Trunculariopsis trunculus,* and *Thais haemastoma.* The next slide [see Plate 9] shows the shells of these three species. In one of the body structures of these beautiful shelled snails, the mantle, there is a hypobranchial gland. The needle in the illustration points to this gland in this specimen of *Thais haemastoma.* This chemical factory has numerous functions, producing a mucoid substance that cements particles as they are expelled by the snail, as well as several neurotoxic chemicals used in predation. And it emits a clear fluid, which is the precursor of the dyes. On exposure to the oxygen of the atmosphere, under the action of enzymes and, importantly, sunlight, the fluid changes from whitish to pus-like yellow, then green, and finally blue and purple.[12] Aristotle, a careful observer, and Pliny give good descriptions of the snails and the process of dye extraction.

Here for instance is Aristotle, 2,350 years ago writing in his *Historia Animalium,* Book V:

> There are many species of the purple murex. . . . The "bloom" of the animal is situated between the quasi-liver and the neck, and the co-attachment of these is an intimate one. In color it looks like a white membrane, and this is what people extract; and if it be removed and squeezed it stains your hands with the color of the bloom. . . . Small specimens they break in pieces, shells and all, for it is no easy matter to extract the organ; but in dealing with the

larger ones they first strip off the shell and then abstract the bloom. [13]

Next slide please [see the illustration below]. Here you see some rock murex adhering to a baited basket in a modern reconstruction of the ancient collecting method. As many as 9,000 animals may have been needed to make 1 gram of the dye.

There is an ancient tale concerning the discovery of Tyrian purple, that of Hercules and his dog. Or, if not Hercules, then a shepherd or a Tyrian nymph. The dog's mouth was stained on crushing a sea-snail with its teeth. [See the illustration on the top of page 169.] Given these tales, and Aristotle's account, it is incredible that the biology of the purple trade remained lost for so long. It was rediscovered accidentally in 1856 by Félix Henri de Lacaze-Duthiers, a French zoologist. He noticed a fisherman draw a yellow design on his shirt with a shellfish of the *Thais* species. The fisherman knew the pattern would turn purplish red.[14]

A reconstruction of the ancient method for collecting rock murex, using a baited basket.

A coin from Tyre, third century C.E. At bottom: Hercules's dog biting a snail.

It should be said that folk knowledge of these snail pigments was always there. And species producing a purple were native to the New World and the Far East as well, so that South and Central American cultures also used the snail purple, as did Chinese and Japanese.

A century later we know much about these pigments. The dyes are the molecules indigo (sometimes called indigotin) and a brominated relative (there is a good bit of bromine in the sea!), 6,6'-dibromoindigo. The chemical structure of the latter is shown in my next slide [see the illustration below].

The chemical structure of dibromoindigo. Normal indigo has the two bromines replaced by hydrogens.

The brominated molecule is Tyrian purple. *Tekhelet,* the biblical blue, is probably some mixture of the two compounds.[15] We don't know for sure, because there is no authentic Hebrew textile dyed in *tekhelet* that has survived. There is substantial variation in the color obtained from the mollusks, depending on the species, the climatic conditions, and the method of processing the dye. Remarkably, even the sex of the specimens matters; for the rock murex *T. trunculus* Ehud Spanier and Otto Elsner find that the male secretes a liquid yielding mainly indigo, but the female's secretion differs, leading to the brominated purple dye.[16,17,18] Just to make life more interesting, some of the snail species change sex from time to time.

While exposure of the snail gland liquid leads to indigo, the dye is not very well absorbed by the wool. So the ancient Tyrian and Hebrew artisans invented some chemistry. [See Plate 10 for a contemporary re-creation.] First, there is a treatment, called "reduction" by chemists, by which the precious snail extract is rendered colorless. In that colorless form the transformed indigo is absorbed well into the molecular structure of the wool, which is a protein. With a little more chemistry to change the acidity of the dyeing solution, the indigo, securely bonded in the wool, on simple exposure to the oxygen of the atmosphere, turns quickly, almost magically, to the desired blue.[19]

Think of the existential agony of the protochemist here—to have gained the colored dye with so much labor, then to be forced to watch its color disappear, hoping, hoping that it will come back!

[*At this point, quite conveniently, the theater's warning bell rings once.*]

I guess this is a sign that I have talked too much chemistry. . . . So let me rush to the present. It took over three millennia from the skillful protochemistry of Mediterranean dye makers, put into the hands of Korah's family on stage, to arrive at the chemical synthesis of indigo.

The synthesis of indigo in the laboratory by one of the great German chemists of the nineteenth century, Adolf von Baeyer, and by Karl Heumann, was in time translated into the crowning masterpiece of the German dyestuffs industry.[20] Today indigo is made synthetically in sufficient quantity to dye about a billion pairs of blue denim jeans a year. The dyeing of each pair consumes 3 to 12 grams of indigo. Then you wash for a long time to get part of the dye out in prewashed designer jeans.[21] We've come a long way (or have we?) from a rebellion in the biblical tribe of Levis.

There is a remarkable book I must recommend to you, *The Royal Purple and the Biblical Blue*. The book's centerpiece is an unpublished 1913 D. Litt. thesis from London University. The author of that work was Rabbi Isaac Halevy Herzog, a photograph of whom is shown in my next slide [see the illustration below]. Herzog was a broadly educated scholar who in his time was Chief Rabbi of Ireland and later Israel. Herzog's thesis, entitled "Hebrew Porphyrology," is accompanied by six diverse papers.[22]

Adroitly weaving his way through the classical descriptions of the dye-producing snails, in full command of the sometimes contradictory talmudic references, aware of the complex zoology of the

Rabbi Isaac Halevy Herzog (1888–1959).

snails, Rabbi Herzog achieves in his 1913 thesis (published merely 74 years later; *that* should be an inspiration to all doctoral candidates . . .) nothing less than a true synthesis. The poetic content of the material he presents presses through, for instance in the variant attribution by the great Franco-Jewish commentator Rashi of the color of *tekhelet* as green, as leeks (Rashi uses the Hebrew transliteration of the French *poireau* to describe the color; though elsewhere Rashi calls it as blue as the sky at dusk). Or the quotation (in eight recensions!) of a saying by Rabbi Meir:

> Why has *tekhelet* been singled out from all other colors? Because *tekhelet* is like unto the sea and the sea is like unto the sky and the sky is like unto the sapphire, and the sapphire is like unto the Throne of Glory. . . .

All along, there has been another, much more economical, source of the same dye. It is the genus *Indigofera* of herbs of the pea family, widely dispersed in warm climates. This plant was an important product of the India trade, for it is readily cultivated there.

Incidentally, even in talmudic times there was persistent trouble with unscrupulous merchants substituting the vegetable dye (just as good a color, but forbidden for ritual use) for the snail dye. The truth-in-labeling problem for *tekhelet* in fact became archetypical of a class of frauds (false weights and measures were another) judged particularly abhorrent in the eyes of God because most consumers could not discriminate between the authentic and the fraudulent.

[The bell rings insistently, twice.]

There is still another source of indigo, and that is the woad plant, *Isatis tinctoria.* It is widespread in Europe and across to Asia. The plant was widely used in northern climates until it was displaced by the southern indigo plant of the East India trade. You may have heard the ancient characterization by Pliny of the encounter of the Roman legions with blue-dyed Celts in 44 and 45 C.E.:

> *Omnes vero se Britanni vitro inficiunt, quod caeruleum efficit colorem, atque horribiliores sunt in pugna aspectu.* (All Britons dye themselves with woad which make them blue, in order that in battle their appearance be more terrible.)

Here is a recipe for a dark blue dye from an Egyptian source, *Papyrus Holmiensis,* of approximately the second or third century C.E.:

> Cut down the woad and place it all together in a basket in the shade. Break the stems into pieces and let them lie a whole day. On the following day let the air in by trampling them, so that the pieces are thrown up by your feet as they move, and dry thoroughly. Put it in baskets and put them in store. Woad treated in this way is known as "kol."
>
> Take about 1 talent (around 55 lb or 25 kilos) of woad and place it in the sun in a tank with a capacity of at least 15 metren (c. 150 gallons, or 600 litres) and pack it tightly. Then pour urine in until it cover the woad and let this heat in the sun. Next day trample round in the woad in the sun so that all is well soaked. This must be done during three days. . . .[23]

The urine is essential. Initially this natural product is slightly acid, but allowed to stand for some time it eventually produces the ammonia whose alkalinity allows the "reduction" to proceed. The instructions given in many sources as to the quantities of urine suitable for a dye-bath contain one or two cryptic modifications: first, the requirement that it must be *human* urine, not only that, but, also, it must come from men, of which the most suitable is that from men who, as the Swede Cajsa Warg puts it in her book of 1762, "drink strong drinks."[24]

So a blue dye can come from a pea-family plant in warm climates, or the urine-processed woad of colder climes. It's a pea or a pee, ladies and gentlemen. [*Audience groans.*]

And what, incidentally, is a pea plant doing making a molecule identical to that produced by a snail? That problem remains to be solved, but surely it has to do with the common biochemical pathways shared by living organisms, and the wondrous games evolution plays. For other examples of a plant or animal surprisingly making the same complex molecule as a very different species makes, I would point to nepetalactone, the active principle of catnip, which comes from a plant (the mint) as well as from an animal (the "walking stick" insect).[25]

[*The bell rings three times; the audience stirs.*]

Ladies and gentlemen, there is precedent for my penchant for digression: Nahmanides. During the Disputation at Barcelona (1276), the fairest of Jewish-Christian debates of medieval times, this great sage of our people takes a page to answer a simple question, where a three-word reply would do. Fray Pul, his opponent, bursts out— and this is reported by Nahmanides—"It is his [Nahmanides's] constant habit to deliver such long-winded speeches."[26]

In fact, we have come to just the right point for me to stop, for it's the actors, and not I, who will tell you of another curiosity in this tale of the Blue and the Purple. In 1887 a remarkable person, Rabbi Gershon Hanokh Leiner of Radzyn, claimed that he had rediscovered the biblical *tekhelet*. We are ready for Act II.

ACT II

SCENE 1. *A street in Naples, 1887, near the harbor, a Naples dirtier and smellier than today, though nary a Vespa scooter is around. But no stage can be as bad as Naples was; indeed this scene shares the features of a set at the nearby San Carlo Opera, and is peopled by some refugees from a production there. Thus the street urchins break into a tarantella at the slightest provocation. The din is incredible—drunken sailors shout obscenities that luckily we can't understand, women with baskets of fish hawk their wares, crippled beggars, pimps.*

From stage left enter four Hasidim, *fully attired in black caftans. They are all sweating profusely, for it is summer, and they are far from their Polish-Russian home. One of them, while young, is wearing a finer outfit, with a* streiml *on his head. He is clearly their leader. The* Hasidim *are accompanied by a nervous, officious merchant, Spacca.*

The entrance of the Hasidim *is such a singularly disconcerting act to the people on the stage that for the only time in living memory in Naples there is a moment of absolute silence, followed by the clearing of a space around them, and whispering.*

Fish Vendor: [*crossing herself*]: Who are they?

Street Urchin: They speak German, I hear that in the north of our country, in Milano, where there are barbarians left.

Knife Grinder: Maybe they're from our lost churches in the East. We should kneel.

Street Urchin: Nah, if it's Spacca who is takin' them around, then it's them who needs protection. He could sell a dead mouse as a duck!

[*In fact the* Hasidim *speak Yiddish with each other, while their* rebbe *speaks excellent French with Spacca. All languages (with the exception of the juicy parts of the sailors' dialogue) are translated simultaneously into English, and displayed electronically in a strip running along the proscenium.*]

Hasid 1: *Rebbe,* why have we come to this strange land?

Hasid 2: *Rebbe,* it is hot.

Rabbi Gershon Hanokh Leiner, the Radzyner Rebbe: Jews, be patient. Motl, Reb Zalman—we have come to Naples, as you know, in search of the *hillazon,* the source of *tekhelet,* our blue. We shall, God willing, renew the blessing of seeing the blue strand in our *tzitzit.* You can take off your coats. . . .

Hasid 3: [*The street children crowd around him; they pull curiously the all-white fringes hanging out from under his shirt.*]: But *rebbe,* how will we know where to find the *hillazon?*

Leiner: Trust me, Reb Levi. I have studied the books of the gentiles as well as I have studied our own. I have learned medicine. I have built our mill that turns out 40,000 kilos of wheat a day, and I have made the machines inside it. In Germany there is a new science called chemistry; even in Russia we have it—professors Butlerov and Mendeleyev have learned it. We will find the *hillazon* and use chemistry to return the *tekhelet* to our people.

[*to Spacca*] Signor Spacca, you have said you will take us to a source of the snail that colors blue, and colors fast, all things.

Spacca: Monsignor, I mean Your Excellency . . . , Your Holiness. Yes. Trust me. Yes, the moment I saw your notice in the paper offering a reward for finding the shell, I knew I would lead you to it. Come look in our famous aquarium, the best in the world. I will show you the beast, and only I know where to find more of them.

Leiner: It must be like a fish, and yet have a shell . . .

Spacca: It lives in the sea, it's home is a shell . . .

Leiner: . . . and have the color of the sea . . .

Spacca: . . . blue as the sea, indeed . . .

Leiner: It must come up every seventy years, but some say seven. . . .

Spacca: So they say, precisely every seven years, but sometimes seventy.

Leiner: Show me the creature, and I will pay you well.

[*They rush to the aquarium at right, the* Hasidim *sweating, the street urchins, braver now, pulling at the fringes of all the* Hasidim.]

SCENE 2. *It is the summer of 1913, a year before World War I. In a dark study, furnished plush red, in the Rue des Francs Bourgeois in Paris, a young man of 25 with a skullcap is sitting at one desk. This is Rabbi Isaac Halevy Herzog, born in Poland, raised in Leeds, England, since age 9, moved to Paris at age 23 when his father Joel was appointed Chief Rabbi of Paris. Joel sits at a grander desk in the same study. Isaac Herzog speaks 12 languages. He is writing his D. Litt. thesis, and is holding up with some puzzlement a letter in his hand.*

Isaac Herzog: Father . . . I do not wish to disturb you. . . .

Joel Herzog: Yes, son.

Isaac: You know the Radzyn *tekhelet* story?

Joel: Yes, Isaac. Gershon Hanokh Leiner of blessed memory was a brilliant scholar. Do you know that at age 23 he began writing a *Gemara* to the sixth order of the Mishnah, something that was missing from both of our Talmuds? But on *tekhelet,* he went astray. Yes, you with your thesis, you know this better than I.

Isaac: Leiner thought the *tekhelet* came from the cuttlefish *Sepia officinalis.* When he came to Naples and saw the creature, he became convinced that this was the seasnail of which our sages of blessed memory spoke. He made a contract for the cuttlefish ink glands;

they shipped them to Radzyn twice a year. There his *Hasidim* did some chemistry, and sold the dye around the world.

Joel: But few wear it. I will not speak of the opponents of the *Hasidim* among our people. But even the other *hasidic* leaders scoffed at Leiner. The *Kabbalah* says that from the destruction of the Temple to the coming of the Messiah, the fringes must be white. To accept Leiner's claim would be to deny their *kabbalistic* text.

Isaac: But I must say something of Leiner in my book. And here's my problem. You know that I have been in correspondence with the great German dye chemist, Professor Friedlander in Darmstadt. In a letter to him, in passing, I made mention of the Radzyn *tekhelet*. Friedlander asked me to procure a sample, and the Radzyn dyeing recipe. This I actually had, because I had written to Radzyn, to the son and successor of Rabbi Gershon. I sent off the threads and the description of their manufacture to Professor Friedlander. He answered [*here Isaac picks up another piece of paper*]:

> . . . As for the dyeing description from Radzyn, what is missing in them is the main thing, which is the dye with which the wool is colored. This is an ordinary modern dye. The specifications of the use of *Sepia off.* are only deceptive, and it is entirely impossible to obtain the coloration in this way.[27]

I rubbed my eyes on reading these lines, father. So I sent the sample to the *Fabrique des Gobelins*. M. Vallette there wrote two months ago that the dye in the sample was likely Prussian blue. I wrote again to Professor Friedlander to ask him if his initial verdict was based on a chemical analysis, for that was unclear in his previous letter. He wrote, in a somewhat peeved tone, I must say, that now he had investigated the sample. He was sure, as a Herr Professor Doktor is sure, that the wool was colored by a mixture of two modern dyes. He suggested I submit the sample to Professor Green in Leeds, the Head of the Department of Tinctorial Chemistry and Dyeing there. And here is his answer, just arrived:

> Dear Sir:
> We have tested the sample of wool furnished and find that it is dyed with Prussian blue.[28]

What shall I do, father? Was Leiner a forger?

Joel: Gershon Hanokh Leiner was a sage, a learned man. In another time, in our time, had he not grown up in Radzyn, he himself would be a chemist, as great as Friedlander, as Green, as great as Adolf von Baeyer, who was Jewish, as great as this young man Chaim Weizmann we hear of. Leiner would not cheat, my son.

Isaac: I will not condemn Rabbi Leiner of chicanery, though I am more suspicious than you, father. Probably he was the victim of his Italian supplier. I'm guessing, and I hope that I am right, that some clever specialist in tinctorial chemistry, in conspiracy with Leiner's agent, Signor Spacca, probably spiked the *Sepia officinalis* ink sacks with synthetic Prussian blue. Someone pulled the wool over Reb Leiner's eyes. Poor man.

SCENE 3. *A brightly lit laboratory at the Shenkar College of Textile Technology and Fashion in Ramat-Gan, Israel, around 1988. All the glassware of* The Man in the White Suit *and* The Nutty Professor *has been brought up to date with the shiny trappings of modern chemistry. And instead of Alec Guinness or Jerry Lewis or Eddie Murphy we have two real chemists—an observant and serious young man with a beard and a prominent forehead, and a feisty older woman with henna-colored hair, rings flashing. It is clear that she is a rebel. Both chemists are wearing laboratory coats, Moshe's pristine, Cora's with lots of acid burn holes and grease stains. Moshe sits at his desk, while Cora with some impatience shuts off a distillation apparatus that has a turquoise liquid dripping through it. She then turns to make some standard laboratory coffee from water boiling in an Erlenmeyer flask over a Bunsen burner. She carries her beaker over to Moshe's desk. [The illustration on page 179 shows the laboratory conditions at Shenkar today.]*

Cora: I'll never get that reaction to go!

Moshe: Oh, you will, Cora.

Cora: [*fidgeting*]: What are you reading?

Moshe: [*He shows her a copy of a blue book with a purple edge around it.*]: It's Rabbi Herzog's thesis on *tekhelet*, finally published. You remember the story.

A contemporary laboratory at the Shenkar College of Textile Technology and Fashion.

Cora: I certainly do. That reminds me, I have something to show you, Moshe. [*She runs over to her desk, rummages in a giant gold fabric purse, and pulls out a small plastic bag. She opens it; and takes out the several strands of blue yarn, blue as the sky, from it.*] Look what I found in Meah Shearim!

Moshe: And what were you doing there? I'm surprised they let you in. [*He makes a vague gesture at her bare arms, her decolletage.*]

Cora: I was with a friend.

Moshe: Let me guess which one: David, Abbie, Aaron. . . .

Cora: This one you haven't met yet. . . . We were going to visit his mother, who is not happy with him since he left the *Hasidim*—and yes, I had long sleeves on, wore a long peasant skirt, and wore thick stockings[29]; I'm not stupid, and I did want to impress his mother. Anyway we were walking along . . .

Moshe: Aha, so it's serious!

Cora: . . . when this kid with earlocks runs after Joseph and says: "Mister, do you want to buy some fringes, real *tekhelet*?" So we buy them. Here is what it says [*she holds up the insert in the plastic bag, while Moshe examines the thread, carefully running it through his fingers*]: "*Tekhelet* of *hillazon*, prepared in the authentic manner of Rabbenu Gershon Hanokh Leiner of blessed memory." It's signed "The Radzyn Community."

Moshe: Yes, they still make it.

Cora: And it's still Prussian blue. Those crooks!

Moshe: Well, Rabbi Herzog of blessed memory didn't say the Radzyner *Hasidim* cheated. He blames it on the Italian supplier.

Cora: Well, that's pretty typical. One rabbi isn't going to badmouth another, no. For that matter, no Jew could do wrong. Of course, it must be the *goyim*. Those rabbis just didn't live in Israel long enough.

Moshe: Take it easy, friend. Herzog is a very thorough fellow and in preparing his thesis he wrote to the Radzyn *Hasidim* to ask them for their dye-making recipe. And they sent it to him!

Cora: [*suddenly interested, grabs the book from Moshe, and reads*]:

> May life and peace together cleave to your honour. . . .
> After enquiring after your honour's welfare!
> Your letter of 20th January [1913] that was addressed to our reverend Rabbi indeed arrived, and since he is temporarily not at home, having travelled from here some days ago, and since I am able to fulfill your request, I hereby inform you that the packet of *tekhelet* enclosed herewith reaches you by post—threads, doubly-sealed as prescribed by our holy law. And in accordance with your request, you can be assured the *tekhelet* herewith enclosed is from the extract of the shellfish's blood without any other dye, and, as I know this work well, I can explain here all the chemicals that

are mixed with the shellfish blood. And whoever is familiar with the
nature of the chemical knows that they are all completely colour-
less. They are all white, and some have the appearance of ordinary
water, except for the blood which is coloured. The shellfish's blood,
that is located in a sac, is as black as ink, and is so coagulated that
it may break and shatter being so hard. This blood is placed in a
very thick vat, and one adds iron filings and also a snow-white
chemical called 'Potasz'. And after keeping on a large, powerful fire
for some four or five hours, until the conflagration burns outside
and within as the fire of Hell, then the blood, the iron filings and
the white chemical fuse and become one piece. . . .

He certainly goes on at some length, doesn't he, Moshe? Let me
skip just a little . . . [*she continues*]:

Afterwards it is emptied into cups, and many encrustations that
adhere to the walls and edges of the cups are filtered and poured
from the cups, and fine encrustations are obtained that are the
essence and extract of the blood. The appearance of this extract
resembles coarse salt on a cloudy day (pale white) 'dunkel Weiss.'
Part of this salt is placed in a pure vat with a quantity of clear water
and the chemicals 'Tartar. deporat. Amoni muriadic. Acid Sulfaric.'
And one places also the wool spun for the *tzitzit* and for the ritual
tekhelet, and a fire is lit beneath the vat. Thus it stays until the wool
absorbs the dye from the vat, and the water remains in the vat.
Again one waits the required time and then one prepares another
vat full of clear water and pours a little 'Acid Sulfaricum' chemical
into the water. A fire is lit under the vat until it is hot, and the dyed
wool is again placed in it and kept there for the required interval,
thereby terminating its dyeing. For, after the wool had absorbed the
dye in the vat containing the shellfish blood and chemicals, since it
tends to be absorbed, the proper appearance was not externalized
on the threads, it being contained and concealed within. But in this
last vat of hot water, the true hue appears also on the outside, just
as it existed already hidden within the wool.
   This is the complete method for making the dye and dyeing the
wool for the ritual *tekhelet*, there being no more. . . . Not knowing
you but with best wishes, Joshua Meir Keitelgisser, writing in the
house of the reverend Rabbi.[30]

Moshe: Note, Cora, that a letter from Paris to Radzyn took just six days to get there. I bet it takes twice as long today. Snail mail.

Cora: This is really interesting, Moshe. Maybe we can figure out what happened in Radzyn.

Moshe: Well, as Herzog guesses, the Prussian blue was probably in there to begin with.

Cora: No, Moshe, you are really dull; get yourself a different *kippah*, maybe with a new skullcap your head will function better. Here we have a laboratory procedure described in some detail. O.K., O.K., so the guy who wrote this didn't take a course with our tyrannical professor, Dr. Zweifler. But the text is telling us something.

Moshe: Alright, let's see what he says. [*He walks over to a blackboard.*] Read again the ingredients.

Cora: They take the cuttlefish blood—Herzog puts a *sic* after that, he knows it's ink and not blood. They add iron filings, "potasz"—that's written out in Latin letters in the Hebrew text—we know what that is.

Moshe: [*writing*] Potash is potassium hydroxide [*he writes ink + Fe + KOH*], pot ash, alkali. And. . . .

Cora: They heat it "until the conflagration burns outside and within as the fires of Hell. . . ." [*Moshe writes a Greek delta above an arrow.*] Let's see, is that it? Yes, that's it. They filter it, evaporate it. And get crystals that look like "coarse salt on a cloudy day." A real poet, Mr. Keitelgisser . . . and he also has a good name for a chemist.

Moshe: Didn't they use some other chemicals?

Cora: Yes. [*She reads*]: Tartar deporat. . . . Probably tartrate. Also ammonium chloride, good old alchemist's sal ammoniac. And sulfuric acid.

Moshe: But all those are mordants, to prepare the wool for taking up the pigment. They have nothing to do with the dye as such. [*He notices that Cora is silent.*]

Cora: Moshe, I've got an idea, this woman has an idea. Moshe, what is Prussian blue?

Moshe: It's a complicated story, I think. Let me look it up in the Gmelin handbook. [*He reaches for a book on the shelf.*] Here's the *Handbuch der anorganischen Chemie. Eisen.* Hmmm. What is it called in German? Oh, yes, *Preussisch blau.* Here it is, *oy vey*—fifty-two pages on *Preussisch blau.* Someone's sure spilled a lot of ink over this subject. . . .

Cora: Spare me the details please; didn't I just see a paper about it in *Inorganic Chemistry*? [*She marches over to the bookshelves and pulls out an issue of the journal* Inorganic Chemistry.] Yes, here it is, by Ludi and Schwarzenbach.[31] Ludi's been working on this structure for many years.

[*She draws the formula KFeFe(CN)⁶ on the board, and begins to sketch the three-dimensional structure that appears below.*]

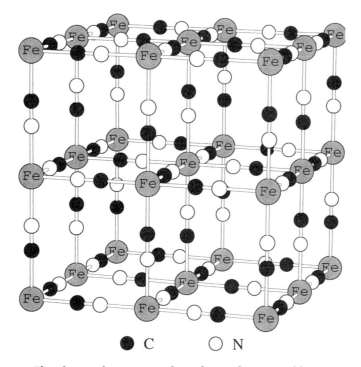

The chemical structure of one form of Prussian blue.

Now I remember the story—there are potassium nonstoichiometries, some water in the structure, a crystallographic nightmare. People knew it was a mixed valence ferroferricyanide, but it took a really good crystallographer to solve the structure. [*She comes out of her chemical reverie.*] Moshe, I've got this idea. I got an idea for how the Prussian blue got into the *Hasidim's* pot. I don't like it, but it's so simple, it must be right. Look at the formula, Moshe.

Moshe: I'm looking, I'm looking.

Cora: What do you see?

Moshe: I see potassium ferroferricyanide. The same thing you see.

Cora: Yes, you dummy. And the Radzyn reaction mixture contains . . .

Moshe: Well, there's potassium hydroxide and iron. And squid ink.

Cora: So we have the K and the Fe (a little oxidation along the way) of the Prussian blue.

Moshe: Yes, sure. But we don't have the cyanide. And Prussian blue is inorganic.

Cora: How did they ever give you that degree at Bar Ilan? Organic, shmorganic . . . Wöhler and Kolbe showed about 150 years ago that you can make organic molecules from inorganic ones, and vice versa. What are you, some kind of vitalist who thinks organic molecules are different because they got soul? And the cyanide, you ask where is the cyanide. Well, what's squid's ink?

Moshe: How should I know? I'm not a biologist. But the squid makes it, in his sleep. So it's organic. Who knows what makes it so dark?

Cora: Yes, yes. . . .

Moshe: So it's probably some protein, maybe some porphyrins.

Cora: And what are those made out of?

Moshe: You know, Cora. . . .

Cora: But I want to hear you say it!

Moshe: Carbon, hydrogen, oxygen, nitrogen . . . [he slows down], maybe a little sulfur. . . .

Cora: Stop right there, *habibi.* So here, my dear friend, is this organic stew, thoroughly rotted in the Neapolitan sun. And it has C and N in it, in the amino acids of the proteins of the cuttlefish. They ship it to Radzyn, laughing all the way at the Jews buying this foul-smelling mess. In Radzyn the *Hasidim* heat the hell out of it, in strong alkali. What if their chemistry, those clever *Hasidim* of Radzyn, *makes* cyanide, CN^-?

Moshe: Cora, I think you're on to something.

Cora: You bet your tush[32] I am.

Moshe: [*leafing through the Gmelin handbook*]: Cora, here is how Prussian blue was made for the first time in 1704. They took some bullock's blood and treated it with potash and iron.[33] The bullock's blood provided the carbon and nitrogen, just like the cuttlefish ink did here!

Cora: Yes, our rabbi wasn't a charlatan. He was a good synthetic chemist. He *made* Prussian blue.[34] So he didn't make *tekhelet.* Perkin made mauve, the first aniline dye, thinking he was making quinine. It doesn't matter—those *Hasidim* did some good chemistry. Moshe, this demands a celebration. Where's that wine you've been hiding?

Moshe: Cora—it's old.

Cora: But it's good. [*Moshe gets the bottle of wine. He pours it into brand new lab flasks, and they drink.*] *L'hayim!* May we always drink old wine in new flasks!

Moshe: [*getting excited all of a sudden*]: Let's check it in the lab, reconstruct the process. . . .

Cora: Right. Now you, Moshe, go down to the docks and get us some cute little cuttlefish. And since it's not kosher, be sure that they don't think you're going to eat it. . . .

[*Curtain descends while Cora gives orders.*]

INTERMEZZO 2. *The audience, as it comes out of the theater, is less subdued, of course, since the end of Cora is very different from that of Korah. It is clear that some of the theatergoers are worried they might be exposed to another lecture, so these (few) head right out to the champagne and* zakuski *snacks. Most people, however, mill about, waiting, and sure enough there is the expected clap of theatrical thunder, and on stage appears another small, bald man. He looks like a twin brother of Professor Azul, but is stylishly dressed in an indigo blue suit, white tie. He begins to speak.*

Professor Marecéu: Ladies and gentlemen: Allow me to introduce myself—I am Professor Marecéu, and I also teach at the Half-Shut University. You have seen in Act II the playing out of the scientific and ethical aspects of the story of the biblical blue (and the Prussian blue that Rabbi Leiner thought was *tekhelet*), and that it has not diminished its hold on the imagination of human beings in the past century.

The government has kindly asked me to speak to you not only of the ethical but of the spiritual ramifications of what you have seen and heard today, and I'm pleased to do so, but at shorter length than my friend and colleague, Professor Azul.

At first sight, Korah's rebellion is but one in a dismally long chain of dissatisfactions of Israel in the desert. What a miserable and inconstant lot! Fortunate enough to live in an age of real miracles (the only one we might see today is the United States winning the World Cup), the Hebrews continue in their desert peregrination to take every opportunity to deny the God who has taken them out of bondage, and who is leading them, through Moses and Aaron, to the promised land.

From the bitter waters of Marah to Korah's rebellion, and past that, the people waver. At some point in this tale of travails punctuated by denial of God or his appointed leaders, the impartial reader who is not religious begins to feel a desperate impatience with the Israelites. Why does God need them? The answer must be that it is not for humans to know, the same answer given in great poetry to Job from the whirlwind. But to quote a later authority, trying to bridge the East-West gap Hollywood-style, "'tis a puzzlement."

There is another aspect to the *midrashic* analysis of Korah's rebellion. One is impressed by the casuistry of Korah. Indeed he is a clever man. We can see that, oh so clearly, in the seductive logic of his argument with Moses on the cloak of all *tekhelet* vitiating the need for a single strand of *tekhelet*. We see an equally seductive rhetoric in Korah's argument to the community, that all are holy, even more startlingly that all should assume authority in rotation.

I have recently been reading a provocative paper by Roald Hoffmann which I wish to share with you.[35] His interpretation of this aspect of Korah's rebellion is that it draws on the historicity of the *midrashic* period, and in particular on the appeal of Hellenistic philosophy and Athenian democracy. There is a struggle with Hellenism in the Talmud and the *midrash*; the pull of pagan culture, and especially Greek philosophy, must have been strong, especially to so intellectually committed a subculture as the scholars of the various academies. For the first time in the history of Israel, there was an intellectual challenge out there (in contrast to idolatry, simple hedonism and brute power), a challenge of beautiful logical complexity comparable to that of the religious tradition, the latter so intricately and intellectually crafted within Judaism by then. Some succumbed; the sad story of the scholar known in the Talmud as Aher, "the Other," is relevant here; of him I will tell you another day.

Korah speaks with the clever serpent's tongue of the Greek sophist. He is sophisticated, too. And wealthy—he can afford those all *tekhelet* cloaks. And Korah's argument to the people sounds very *democratic*. Nay, "revolutionarily democratic," to quote Dr. Hoffmann. He cites another part of the *midrash* here, to illustrate the short distance between Korah and Che Guevara:

> Korah assembled all the congregation, and in their presence began to speak words of scorn, saying, "In my neighborhood there was a widow, and with her were her two fatherless daughters. The widow had only one field, and when she was about to plow, Moses said to her, 'Thou shalt not plow with an ox and an ass together.'[36] When she was about to sow, Moses said to her, 'Thou shalt not sow thy field with two kinds of seed.'[37] When she was about to reap the harvest and to stack the sheaves, Moses said to her, 'Thou shalt not harvest the gleanings, the overlooked sheaves, and the corners of the field.'[38] When she was about to bring the harvest into the granary, Moses said to her, 'Give heave offering, first tithe, and second tithe.' She submitted to the law and gave them to him. What did the poor woman do then? She sold the field and bought two sheep, so that she might clothe herself in the wool shorn from them and

profit by the production of lambs. As soon as the sheep brought forth their young, Aaron came and said to the widow, 'Give me the firstling males that are born of thy herd and thy flock, for this is what the Holy One said to me: "All the firstling males that are born of thy flock thou shalt sanctify unto the Lord thy God."'[39] Again she submitted to the law and gave the firstling males to Aaron. When the time for shearing arrived, she sheared her two sheep. Then Aaron came again and said to the widow, 'Give me the first portion of the shearing, for the Holy One said, "You shall give him . . . the first shearing of your sheep."'[40] She said, 'There is no strength in me to withstand this man; behold, I will slaughter the sheep and eat them.' After she slaughtered them, Aaron came once more and said to her, 'Give me the shoulder, the jaws, and the maw.' The widow said, 'Even after I have slaughtered my sheep, I am still not free of his demands; behold, I devote my sheep to the Sanctuary.' But Aaron said to her, 'In that case, they belong entirely to me, for thus said the Holy One, "Everything devoted to Israel shall be thine."'[41] Aaron picked up the sheep, went his way, and left her weeping—her and her two daughters. This is what happened to the poor woman. All such [evil] things Moses and Aaron do on their own; but they hang the blame on the Holy One!"[42]

It is true that Greek democracy had but a brief flowering in Athens. Still, the democratic ideal was much talked about. It had a strong hold on the intellectual imagination of the Hellenistic world, and survived into the time of the composition of the *midrashim*.

There was no room for Korah in the ideology of the strongly led Exodus, nor in the politics of the Roman Empire and its system of vassal kings. Korah had to be destroyed. Dr. Hoffmann says that it is startling, however, to see two faces of Moses so close to each other— the Moses who appeals to God for mercy to spare the innocent, lest they die with the guilty, and the Moses who appeals to God to make the guilty perish in such a terrible and abnormal way (to die and for one's complete body not to be found and given decent burial was a terrible fate in the Mediterranean world[43]), in order to establish Moses' authority! Moses was not a simple man.[44]

At this point there will be some in this distinguished audience who will think that I have been unleashed upon you by the remnants of the Soviet Ministry of Anti-Religious Propaganda. This is hardly so, and now I will tell you another view of the Korah rebellion and its central importance. This perspective is drawn from the thought of one of the great Jewish religious scholars of our time, Rav Joseph B. Soloveitchik, of blessed memory.

Here's how the Rav, as Joseph Soloveitchik was called, characterized Korah's uprising:

> The rallying cry which Korah chose was "common sense." He proclaimed that all reasonable people have the right to interpret Jewish law according to their best understanding: "For all the community are holy." In down-to-earth logic, the lowliest woodcutter is the equal of Moses. This appeal to populism evokes considerable support because it promises freedom from centralized authority; it flatters the people's common intelligence and it approves the right of each Jew or group of Jews to follow his own individual judgment.[45]

Korah's argument, Soloveitchik points out, is indeed the basis of all "subjectivist" arguments for redefining Judaism as being based *not* on strict religious observance of the precepts (*mitzvot*) agreed on in the compact between God and the Israelites, but rather on some vague notion of personal inner feelings, mood, just a sense of being Jewish.

The particularly insidious nature of Korah's argument was that it combined a reliance on intelligence (it is not logical that . . .) with an appeal to emotions and individual prerogative (all the community are holy). Why do Korah's words sound so powerful and hardly anachronistic today? Because it is just these forces—intelligence (represented arguably by science) and emotion (all the ways we have of indulging body and soul, from the jacuzzi to pornography)—that play a melodious jingle, then gambol and sneak into our souls today. They're a formidable lot, intelligence and emotion together.

A second point the Rav makes was that Korah's immediate challenge was to the performance of the very specific and detailed observances of the religion, in the case at hand to the tying of the fringes and the blue thread among the white. Why use a blue thread if the whole blue garment can induce in you a feeling of the sublime? The answer the Rav gives is that, yes, there are two levels of observance, "the objective outer *mitzvah* and the subjective inner experience that accompanies it. . . . The former without the latter is an incomplete act, an imperfect gesture." But the latter is absolutely essential, for it induces the emotion:

The Torah, therefore emphasizes the mitzvah, which reflects God's will; it has the stamp of immutability and universality. The great religious romance of man with God, the emotional transport, follows one's observance of the *mitzvah*, not the reverse.

Moses was unquestionably right. If one fulfills the mitzvah of tzitzit, recognizing its religious meaning, then a glance at one blue thread will produce an awareness of God. To this day the *tallit* [prayer shawl] (even without the blue thread) is religiously inspiring to the worshipping Jew. Such is the power of the *mitzvah*. Proceeding from action to feeling, the blue color can remind one of his link with God. However, if one fails to conform to the halakhic norms and instead, availing himself of common sense, substitutes a garment that is entirely blue, his response will be divested of its religious meaning and totally secular. And if there is a response at all, it will be a mundane, hedonistic experience, aesthetic appreciation, but not a religious emotion.[46]

There is another point that Rav Joseph Soloveitchik did not make, but which I will, following the article by Roald Hoffmann I mentioned. The relaxation of observance, the appeal to a classless, unmediated communion between human and God—that was exactly what made early Christianity so attractive to some Jews and to non-Jewish multitudes. In time (it did not take long) the Catholic church dropped the "unmediated" part, and, recognizing the deep appeal of ritual to the human soul, instituted its own rich panoply of ritual. But Korah's appeal has parallels in the voice of Christianity in *midrashic* times.

So far little has been said by me of the spiritual meaning of the blue and the white, especially the blue of the sea, the blue of the sky, the blue of the throne of God. Rav Soloveitchik has written another essay on this theme, an essay of great philosophic import and sublime poetic power. He reviews the facts of the dye's origin and disappearance with time, just as my colleague Azul did. And Soloveitchik notes the contentious discussion (which we did not mention, but which is an excellent illustration of the sages' agreeing to disagree[47]) on the ratio of the number of white to blue strands.

But the Rav, in whom the greatest intelligence and religious scholarship were coupled to a temperament that is perhaps best characterized as existentialist, at this point does something incredible. He mines ever so deeply the symbolism of the blue and the white. In our sad loss of the blue, as sad as the expulsion of the Jews from the Iberian peninsula toward the end of the fifteenth century, the Rav sees a metaphor for the human condition, for "the lonely man of faith," as the title of one of his books has it.

The color white is a symbol of rationality and purity. But what of blue? One reaction might be "hope" (or is that quality symbolized by green, as the ever-renewing earth?). But no, to the Rav the blue of *tekhelet*

> is the "likeness of the seas and the heavens" and focuses our thoughts on the grand mysteries of human experience which elude our precise understanding. The seas and heavens are boundless and beyond human reach. They encompass the abstract and the transcendent, ultimate value and ends, man's metaphysical quest and his efforts to rise above the self-evident and the temporal. It is this area which remains a perennial enigma, resisting rationalization and quantification. It is the realm of philosophy and religion which postulates truisms even as the great mystery persists and precise decipherment proves elusive. While the color white bespeaks the clearly perceptible, tekhelet refers to a realm which is only vaguely grasped.[48]

If I continue to quote Rav Soloveitchik, my dear friends, it is only because I do believe that it has never been said better:

> The many threads of white, according to Maimonides, urge us to use our minds for discoveries in technology, to explore and master nature. The universe will yield its secrets to the organized scientific pursuit. But the one thread of tekhelet pertains to the spiritual realm, where man is humbled by the mystery of existence. Here he needs the guidance of revelation and the religious perceptions of the soul. . . .
>
> The same dichotomy between being on terra firma and on shifting sand is also experienced in our personal life. We have all had periods which are rational, planned and predictable, when we feel that we have a hold on events. At other times, however, mystery and puzzlement intervene, dislocating the pattern of our lives and frustrating all our planning. No one can say, "The world and I have always gotten together reasonably, happily, and successfully, with ambitions always being realized. I have never been defeated." Stark and harsh reality often imposes the bizarre and the irrational, leaving us stupefied, shocked, and bereft.

Inexplicable events render us humbled. This is the tekhelet of human experience.[49]

And now [*The bells ring once, then twice, then three times in rapid succession.*] I will let the Rav make the transition to what you will see in the third and final act of "The Flag That Came out of the Blue":

> Only a people sustained by *tekhelet* could be motivated to reconstitute a state after two thousand years of exile. Nations governed only by 'white' mock us incredulously and derisively. We are sustained by tekhelet, even when it is only a vision and temporarily obscured. The garment of Jewish life will yet possess both blue and white, and our historical yearnings and sacrifices will be vindicated.[50]

ACT III

SCENE 1.* *Evening of August 25, 1897. A room in the Hotel Three Kings, Basel, Switzerland. Theodor Herzl, 37, the founder of modern Zionism, is seated facing the audience stage left. He is writing in his diary, stroking his full black beard. A white flag with seven gold stars is draped over a cabinet. There is a knock on the door and David Wolffsohn** enters. Wolffsohn is about the same age, also bearded.*

Herzl: Thank you, Daade, for coming. There is so much for us to do. . . . Tomorrow we write history, the First Zionist Congress will begin.

Wolffsohn: We're expecting 197 delegates from 15 countries—Russia, Germany, Bulgaria, the United States, Algeria, Palestine, et cetera. They represent old "Lovers of Zion" societies, newly formed political Zionist groups, in some cases only themselves.

Herzl: Still I worry about the rotten eggs among them. You remember Bambus, the leader of the Berlin "Lovers of Zion," who turned

---

* Most of the words in this scene are taken from the diaries of Theodor Herzl, *The Complete Diaries of Theodor Herzl*, edited by Raphael Patai, translated by Harry Zohn (New York: Thomas Yoseloff, 1960).

** David Wolffsohn (1856–1914) was a German-Jewish merchant and Zionist leader. He was born in Lithuania, and lived much of his life in Cologne. He became one of Herzl's earliest followers (Herzl affectionately called him Daade), and played an essential role in the workings of the World Zionist Organization. Following Herzl's death, Wolffsohn succeeded him as president, from 1905–1911.

against us. And Dr. Kohn, of the "Group Kohn-Rapoport," also known as group Korah, who tried to bring about a split among the Zionists in Vienna. He's here. I guess we'll have to let them in, Daade, but they are against us.[51]

Do you realize, Wolffsohn, we are laying the foundation stone of the house that is to shelter the Jewish nation, to obtain a publicly recognized, legally secured homeland in Palestine?

Wolffsohn: Theodor, you are the dreamer whose dreams come true. I will follow you, to Palestine and not Argentina.

But let's go over the practical details. How did you like the Burgvogtei hall that was rented for the Congress?

Herzl: Inappropriate—we just can't look ridiculous. That hall has a gaudy vaudeville stage. I've already done something about this, engaging more dignified quarters—the Basel city casino. [See the illustration below.]

Herzl, in white tie and tails, presiding at a Zionist Congress in the municipal casino, Basel.

Wolffsohn: And attire?

Herzl: One of my first ideas, months ago, on how the Congress should be conducted was to have black tails and white tie obligatory at the opening session. Full dress has a way of making most men feel rather stiff. The stiffness induces a measured, deliberate tone—one not easy to come by in light summer suits or travel wear. I will spare nothing to heighten this tone to the pitch of solemnity.

[*At this point a waiter knocks and discreetly brings a supper tray of two soft-boiled eggs and smoked mackerel.*]

Herzl: Danke schoen. But I asked for only *one* boiled egg . . . oh, never mind. I will find a use for the second one. [*He pauses for a moment.*] Daade, there is something that concerns me, and it also has to do with attire.

Wolffsohn: Yes, yes, you do look troubled.

Herzl: In deference to religious considerations, I will go to synagogue on the Saturday before the Congress. The head of the congregation will call me up to the Torah. I had the brother-in-law of my Paris friend . . . drill the *brokhe* [benediction said upon reading from the Torah] into me. I know that when I climb the steps to the altar, I will be more excited than on all the Congress days. The few Hebrew words of the *brokhe* are causing me more anxiety than my welcoming and closing address and the whole direction of the proceedings.[52]

Wolffsohn: We'll review the blessings. . . . As for attire—your formal dress will be just fine; you do need a hat and a prayer shawl. Here—use my *tallit*. [*Wolffsohn takes a large white wool rectangular prayer shawl out of an embroidered bag. It has a blue stripe along each end, and a long wool tassel tzitzit in each corner. Note to costume designer: See pictures of prayer shawl in Chapters 3 and 5 of* Old Wine, New Flasks *by Roald Hoffmann and Shira Leibowitz Schmidt.*]
   This is how you put it on.

[*Demonstrates. As he removes shawl from the bag, he checks each* tzitzit, *untangling the threads to confirm that none is torn or missing. Puts still-folded* tallit *on his shoulder and makes the blessing*]:

Blessed are You, O Lord our God, King of the universe, Who has sanctified us with His commandments, and has commanded us regarding the commandment of *tzitzit.*

[*Takes tallit by edges of neckband, kisses center of neckband, raises his arms, swirls tallit overhead and behind him, his arms lowering the tallit onto his head so it covers his face as he says the blessing thanking God for the commandment that we wrap ourselves in a tallit. With a sweep of his arms he then wraps the full length of the tallit over his head, forming a thick hood. He then lets it drop into a position where his shoulders are draped by the prayer shawl, which hangs down to about 15 inches from the floor.*]

Herzl: I'm reassured, Daade. Strange, here I'm founding a Jewish nation, yet I do not know the customs of our people. As you do.

Wolffsohn: You will learn them, Theodor. This is a new world, and even in the old one, two thousand years ago, the Jewish people's leaders were not the people's high priests.

[*Wolffsohn, still wearing* tallit, *plucks the white flag off the cabinet.*]

And this must be a prototype of the flag you describe in your book. You wrote, I remember almost by heart,

> We have no flag, but we need one. If we want to lead the masses, we must wave a symbol above them. [See the illustration on page 196.] I imagine a white background with seven gold stars. [*Gestures to white and gold*] The white stands for our new and pure life; the seven gold stars are the seven-hour working day. We want to better workers' conditions; we shall enter the Promised Land in the sign of work.

Herzl: Daade, it's touching, you know it by heart. A flag is a potent signal. Here is what I see: As soon as we sight the new shores from the pioneering ship, the flag of our society—which will later become the national flag—will be hoisted.

All will bare their heads. Let us salute our flag.

The first man ashore will carry a cheap, shoddy flag in his hand. It will later be preserved in the National Museum.

The flag Herzl designed did not become the flag of Israel. But Herzl need not fret—thousands of Israelis carry it in their pockets today, on the face of a phonecard marking the centenary of the First Zionist Congress. (Designed by Sharon Murro for Bezeq.)

Wolffsohn: Not all will bare their heads, dear Theodor, not in the old-new land of Israel. . . . You are so far from your roots, from the masses, that you don't realize even this, nor that gold and white are not *our* colors. I grew up in Dorbian, a small *shtetl* in Lithuania and I know what is in the soul of our people. It is not gold. It is blue and white. A special blue, blue as the sea, blue as the sky, blue—some say—as the throne of God. *Tekhelet* and white are our colors. This [*grasps* tallit] is our banner. And the star of David, the king, for whose return we long. And the lion of Judah, our strength. These are our symbols.

Herzl: You remember the whole section about the innovative seven-hour work day in my book. *My* flag proclaims the significance of social progress. . . .[53]

Wolffsohn: But *our* flag has cosmic significance. [*Removes and folds* tallit *lovingly, replacing it in cloth bag.*] Look, you listened to me when I coined the name *sheqels* for the dues we pay the nascent Zionist movement; that was based on the biblical *sheqel* coin. Listen to me now. You are a writer, you have a poet's soul. Listen to these lines by the poet, L. A. Frankl, who wrote some twenty years ago in a Hebrew newspaper,

> All that is sacred will appear in these colors:
> White—as the radiance of great faith
> Blue—like the appearance of the firmament.

[*There is a knock on the door. Herzl opens it to find two young men dressed in pioneer-style work clothes, far from white tie and tails. They lift their caps, speak in accented German.*]

Israel Belkind: Belkind, Abramovitz. We wish to present our credentials as the delegates from Rishon Le-Zion, the new town in Palestine that we built in 1882 on sand dunes south of Jaffa. We are proud to pay our *sheqel* dues.

[*Wolffsohn nods "I told you so" to Herzl.*]

Herzl: [*ever the politician*]: Thank you, thank you, friends. Just a moment, I have something for you. Help me, Daade [*goes to look for something in the adjoining room*].

Zeev Abramovitz: [*to his colleague*]: Look what's hanging there. That must be the flag he's going to propose to the Zionist Congress.

Belkind: [*whispers back*]: He's so out of touch with those who are working the land in Palestine. He's on another planet—doesn't know we have a flag. I remember how . . . when was it? . . . about a dozen years ago . . . you, Zeev, thought we needed a flag for a parade on the third anniversary of our settlement's founding. Your wife embroidered a blue "shield of David" on a white rectangle modeled on your *tallit*, with blue stripes, and your son marched with it in the parade. We should tell him. [See Plates 11 and 12.]

Abramovitz and Belkind: Let's not argue now with Herzl, after all he has done. He talks to the Sultan and the Kaiser; we dig the earth. There is time for a flag.

Herzl: [*returns*]: Here gentlemen, is a volume containing six of my latest plays, all performed in Vienna. You will appreciate the social questions I address, improvement in workers' conditions, the seven-hour work day. Some of the plays have Jewish themes as well.

Abramovitz and Belkind: Thank you, thank you [*nodding in appreciation as they exit*]. *Shalom aleichem.*

[*Door closes. Herzl looks to Wolffsohn for translation.*]

Wolffsohn: That means "Peace be with you." Next time respond *Aleichem shalom* . . . you hopelessly assimilated Viennese Jew, you. . . .

Herzl: And you, my dear Daade, will be the hero of my next novel, *Old-New Land*. I shall call you David Litvak and portray you in glowing terms as a citizen of our new state. You know what endears you to me? You disagree with me on matters that I know insufficiently.

Wolffsohn: . . . like the flag.

Herzl: You talk of flags, when I have so many other delicate problems. [*Walks over to the room service tray with the two boiled eggs, picks up one egg in his left hand, while pantomiming, à la Marcel Marceau, a dance on eggs.*] I am dancing on eggs. . . .

The Egg of my employer, the *New Free Press,* to which I must not furnish a pretext for giving me the sack.

Egg of the Orthodox.

Egg of the Modernists.

Egg of Turkey and the Sultan who rule Palestine now.

Egg of the Russian government, against whom nothing disagreeable may be said, although the deplorable situation of the Jews there must be mentioned.

Egg of the Christian denominations with regard to the Holy places.

In short . . . all the difficulties that I have long been battling. To which may be added a few more eggs. [*In his right hand takes the second egg from its egg cup, brandishing it at each phrase*]:

Egg of Edmond de Rothschild.

Egg of the Lovers of Zion Society in Russia.

Egg of Palestine settlers, like those sweet boys who came here.

Egg of jealousy, egg of envy.

I must conduct the movement impersonally [*tosses egg in air planning to catch it in his left hand*] and yet not allow the reins to slip from my hands [*catches egg but it slips from his hand, cracks on the valet, and being quite a soft-boiled egg, splatters over the white/gold flag prototype*].

Wolffsohn: Great! Now you'll have to have a new flag made up anyway.

[*There is a knock on the door, Herzl opens it.*]

Mrs. Sonnenschein: Good evening. I am Rosa Sonnenschein, and I want to give you a copy of *American Jewess* magazine of which I'm the editor. [*Glances at egg scrambled over the flag.*] Looks like it's just the moment you need a woman here. [*She takes flag to washbasin.*] Do you know, Mr. Herzl, what Israel Zangwill wrote about you?

> A majestic Oriental figure . . . with eyes that brood and glow—you would say he looks like one of the Assyrian Kings, the very profile of Tiglath Pileser. . . . In a congress of impassioned rhetoricians he can remain serene, his voice . . . subdued . . . and yet beneath lurk the romance of the poet and the purposeful vagueness of the modern evolutionist; the fantasy of the Hungarian; the dramatic self-consciousness of the literary artist; the heart of the Jew.

I can see he was right. [*Herzl bows*] But I hear in the hallways plans to dethrone you, Tiglath Pileser. [*Significant pause.*]
They will crucify you yet—and I will be your Magdalene.

Herzl: [*not quite rising to the occasion, with a glance at Wolffsohn*]: Thank you for this magazine. It will help me understand the American delegates better. . . .

Mrs. Sonnenschein: [*pointing to open book on desk*]: That must be your diary. It will make fascinating reading for my grandchildren to read your account of this Congress.

Herzl: They will be able to read about you and your *bon mots* in it. [*She does a playful curtsy.*] You, as an American, will appreciate what a London correspondent who interviewed me here yesterday called my diary: "The Log Book of the *Mayflower.*"

Mrs. Sonnenschein: . . . a *Mayflower* carrying pilgrims back to the original Zion. Good evening, captain. [*Mock salutes, smiles, exits.*]

Herzl: If I were to sum up the Congress in brief—which I shall take care not to publish—it would be this: At Basel I founded the Jewish State. If I said this out loud today, Daade, I would be greeted by universal laughter. In five years perhaps, and certainly in fifty, everyone will perceive it. The essence of a State lies in the will of the people for a State.

Wolffsohn: Let's see, in fifty years, that would bring us to 1947. I wonder if we will have our state by then.[54] And what will we have a century from now, in 1997?

I'm late [*pulls watch on chain out of his pocket*] to the finance committee meeting. I'll let you know how it goes.

Herzl leaning on the railing of the Three Kings Hotel in Basel, 1901, at the Fifth Zionist Congress. (Photograph by Ephraim Moses Lilien.)

Herzl: Goodbye. [*Takes some Swiss francs from wallet.*] Here, for the tailor.

Wolffsohn: But I told you, I already have my tails and white tie ready.

Herzl: Not for that. This is for your flag. White and blue . . . or how did you call that special shade of blue in your Hebrew? *T . . . t . . . Teke* . . . ssssomething. . . .

Wolffsohn: *Te-khe-let.*

SCENE 2. *It is late at night in Basel, on a bridge. Medieval buildings loom to one side, the chemical factories of Ciba-Geigy, Sandoz, and Hoffmann–La Roche to the other. The fog is coming in, the kind of rolling fog that is tangible, clouds close to the ground. Gradually the fog covers most of the stage so that by the end of the scene, the buildings are entirely obscured; nothing is visible of the city. One sees only the bridge, solitary street lights, the waters of the Rhine, and the two actors.*

The date? In that universe in which our characters are about to move, it doesn't matter.

As the scene opens, Theodor Herzl is leaning on the bridge railing alone, gazing into the night, deep in his thoughts. The pose is recognizable as that of his well-known portrait, in which it is a hotel balcony in Basel that he leans on, looking toward our bridge. Yehuda Amichai* walks quickly out onto the bridge. He is still a young man. He wears a trenchcoat. Amichai looks at Herzl; Amichai is puzzled because the bearded man looks familiar to him. He walks by Herzl, who doesn't see him, turns around and stops on the other side of the bridge. Amichai takes out some sheets of paper from a briefcase, scribbles some revisions, all the time glancing at Herzl. He wants to talk to Herzl, but the latter is so deep in thought that he cannot be interrupted.

---

* Yehuda Amichai is acknowledged as Israel's greatest living poet, one of the first to use idiomatic Israeli Hebrew in his reflective and poignant poetry. He was born in Würzburg, Germany in 1924 and immigrated with his family to Palestine in 1935. He served in the British Army and in the Negev Brigade during the Israeli War of Independence. Amichai's numerous volumes of poetry, children's books, plays, and novels have been translated into many languages.

Amichai fidgets, and tries out a few lines of a poem in Hebrew. The sound of the language startles Herzl, whose turn it is now to be mystified by this wild young man speaking Hebrew on a bridge in Basel. Amichai makes a final change in the poem and then reads it out loud. Now he couldn't care whether Herzl were there or not.

While Amichai reads there is projected on a screen (only intermittently visible through the rolling fog) a series of silent images of the story of the Jewish people and Israel from the Zionist Congress to the time when Amichai wrote the poem. The montage may include pictures of pioneers on a kibbutz, old and new Tel Aviv, scenes of Jewish politicians (Leon Blum) and simple people in a market in Zloczow, scenes of the Holocaust, the ship Exodus, the Haganah (Rabin and his comrades) training, the battle for Jerusalem (flags hoisted), the surrender of a fortress on the Suez (the soldier emerging with a Torah in his arms), the Temple Mount, Meah Shearim. Throughout this montage, which moves very rapidly, without lingering, are interspersed flags and scenes of nature—from a hydroponic flower garden, to the sea of Tiberias, to the desert blooming with tomatoes, a vineyard. The dominant colors are in the range of those attributed to tekhelet, from green to blue to reddish purple. It seems the fog is tekhelet-tinged.

Yehuda Amichai:
How did a flag come into being?
Let's assume that in the beginning
there was something whole, like
the dress of a woman
you long for, which was
then torn into two pieces, both big enough
for two battling armies.

Or like the ragged striped fabric
of a beach chair in an abandoned
little garden of my childhood,
flapping in the wind. This
too could be a flag making you arise
to follow it or to weep at its side,
to betray it or to forget.

I don't know. In my wars
no flag-bearer marched in front
of the gray soldiers in clouds of dust and smoke.

I've seen things starting as spring,
ending up with hasty retreat
in pale dunes. I'm now far away from all that, like one
who in the middle of a bridge
forgets both its ends
and remains standing there
bent over the railing
to look into the streaming water.
This too is a flag.[55]

[*As he hears the poem, Herzl moves, slowly. By the time the poem ends the two men, one in full dress coat, hat, and tails, the other in a scroungy raincoat, are facing each other. For an instant it even seems they might hug. But that moment passes, and nothing is said. Amichai walks off. Herzl returns to gaze at the Rhine. The curtain falls.*]

SCENE 3. *The coffee lounge in the lobby of the Basel Hotel in Tel Aviv, Saturday, November 4, 1995, a warm fall evening, 9:45 p.m.. Three young people, a boy and two girls, are sitting, sipping on the traditional Mediterranean coffee, known affectionately as "mud." They are seated at a table for four in the corner, near several Israeli flags. One chair is empty. From time to time they peer through the tall glass window that looks out on a busy Tel Aviv street. It could be a café anywhere in the world, except for the rifle slung across the back of the boy's chair, the* kippah *clipped to his hair and the* tzitzit *peeking out of his shirt, and the girls' jeans skirts brushing the floor. Geula Abramovitz is leafing through a copy of* Nekuda, *the magazine of the West Bank settlers' movement. The other girl (Rachel Osher) is flipping the pages (with not much interest, to be sure) of a chemistry textbook. Aviram Reuben is absorbed in a book by Yehuda Amichai. From time to time in the ensuing conversation, they look toward the door—it seems clear that they are waiting for someone.* [56,57]

Geula Abramovitz: [*shaking her head*]: Did you see what our politicians want to do next [*she slams the magazine down*]!? To give back Bethlehem *ahead* of the Oslo agreement schedule! And not only Bethlehem; every city in Judaea and Samaria.

Aviram Reuben: And the Golan too. You heard our friend [*pointing to empty chair*] quote that passage in the Talmud forbidding the return of the cities at the borders? I reread it after our last talk and I see how fraught with danger, physical danger, the sages felt such a policy was. I really don't know what should be done.

Rachel Osher: Why can't we trade the land for peace?

Geula: We've been over that, Rachel. Because this land, won through our blood, is the beginning of the redemption. Getting back the Greater Israel, the Israel of Solomon and David, prepares the ground for the footsteps of the Messiah.

Rachel: So you know when he is coming, do you? I remember a passage in the Talmud when Rav Zera came upon some scholars who were busy calculating the date that the Messiah would come. And he said to them—don't postpone the coming through your calculations and expectations, for three things come unawares: the Messiah, a lost article that you are searching for, and a scorpion.

Geula: Don't quote Talmud at us, Rachel. Or is it Rochelle? You Americans didn't fight for the land as we did. And I'm not the only one who feels this way. From Rabbi Abraham Isaac Kook, of blessed memory, on we've felt that this struggle for a state is a sign. It's not an accident my parents named me Geula. I was born after the Six Day War, when it became even clearer that this was "the beginning of the redemption": *at'halta diGeula.* And I'm not the only one. Aviram's friend [*nodding at empty seat*] was also named for the beginning of "redemption." Our names didn't come out of the blue! Just as our flag didn't come out of the blue. . . . Well, actually, maybe the flag DID come out of the blue, out of the *tekhelet* thread that once was in the fringes you're wearing. [*She points first to the flag in the lobby and then to Aviram's fringes*] The redemption, Geula, and the flag grew from the Bible. Our core and essence are rooted in the land. In this land!

Aviram: The funny thing is that Rabbi Kook's ideas fitted well into the ideology of those socialists who ran our country at the beginning. The ones who let Kibbutz Mizra grow pigs and sell pork sausages.[58]

Rachel: How do you explain that?

Geula: For Rav Kook the value of settling the land outweighed all else—even the laws on prohibited food. There was an analogy: the nonreligious, even antireligious, pioneers were like the nonpriestly laymen who were needed to construct the Holy of Holies in the ancient Temple. Ordinarily forbidden to enter that sacred space, they were nevertheless allowed to do so to prepare the way for the Priests. The holiness of the land of Israel, the nation of Israel, the Torah were all three inextricably tied with each other in Rav Kook's theology.

Aviram: Now that you started talking about food, I realize I didn't have any supper. Do you mind if I order something substantial? [*He signals waitress, who approaches.*] Can I have two eggs, and toast? Thanks.

Geula: And Ramban counted the settling of the land among the 613 *mitzvot.* So that is one major source of authority. Another is Rav Kook's son, Rav Zvi Yehuda, who applied the teachings of his father to the new situation—when we won the Six Day War—and taught that the *halakhah* says we must hold on to Greater Israel. *Tekhelet* and white must fly over Judaea and Samaria!

Rachel: Do you, Geula, ever try to see the other side of an issue? You say Nahmanides counted settling the land in his enumeration of the *mitzvot.* But the Rambam, Maimonides, did not count settling as a *mitzvah.* . . . And for many, perhaps most, the authority of Maimonides is decisive.

Geula: Still, the opinion of Nahmanides is legitimate.

Aviram: We aren't talking of legitimacy! You are both right, or right and left [*smiles indulgently*]. It's a problem of emphasis. Maimonides was the great rationalist, whereas Ramban—Nahmanides—had mystical leanings and therefore perhaps saw a special light in the land.[59] And in our century Rav Kook was very lenient in cooperating with those who were settling the land. But the flip side is that maybe Rav Kook, and later his son, exaggerated the aspect of holiness in the land.

Rachel: [*looking at Geula*]: Or maybe you and your teachers want to use the great Rabbi Kook for your own purposes. This is what I've heard Rabbi Carmy say in a lecture at college—that in your propaganda he doesn't recognize the Rabbi Kook he has studied, that you've censored and misinterpreted him, that you've just used him as a way of outzionizing the Zionists.[60]

Geula: But the land is holy!

Rachel: Is it? Do you think the molecules of earth, that [*thumbing through the chemistry text*] nitrates, silicates, ferrugineous clays [*that last term she spews out with a confident flourish*], that the chlorophyll in the leaves are intrinsically holy? Show me the molecule that has the holiness!

Look what happened to Korah, who was smugly sure he was already holy. Compare Korah's slogan with God's command in the section just before the Korah episode. Korah says: " . . . for all the community *are* holy." And God says: "You are *to be* holy." God presents an aspiration, something for humans to strive for. But Korah is sure. Damned sure, as it turns out.

Aviram: Korah was driven by ambition, and was a man of duplicitous rhetoric; he was not committed to an ideal. So why bring in Korah?

You can't stop to examine the merits of every side when your life is threatened, Rachel. That's why we went to the army *yeshivah-hesder* program. You underestimate the real physical threat to the nation, Rachel [*he adds, enjoying his role as arbiter*] . . . and maybe Geula overestimates it.

[*Pauses; the soft-boiled eggs in their egg cups and toast come.*] Let me go wash. [*Goes to a sink and does the ritual handwashing before bread, returns, says the blessings, and bites into the toast.*]

The people of Israel have lost their ideals. Take the army: Once the *kibbutzniks*, like those of the socialist collective farm Mizra that you mentioned, were the major source of officers. Over the decades they got tired, because their commitment was built on ideas and not faith. When they died out, and everyone tried to get out of the army and into Harvard Business School, we were there! And what's happening now is that more and more of my *yeshivah* friends are rising in the military.

Geula: Why do you think I signed up for the territories? The settlers in Judaea and Samaria are the "pioneers" of today, we've replaced the old *kibbutzniks*, in soldiering and in settling. [*looking at her watch*] I wish I knew where our friend is.

Let me try to explain another way, Rochelle. My grandfather told me his father, Zeev Abramovitz, left his family in Bessarabia to build the town of Rishon Le-Zion—the "first in Zion." There was nothing there, just some sand dunes. Grandfather says when he was a boy he carried the first blue-and-white flag: It was a star of David added by his father to his blue-and-white prayer shawl for a town parade.[61]

Aviram: So that's where you get your bold blood [*smiles a little at her*].

[*At this moment another student, Sasha, enters, sees the group and comes over. His excellent Hebrew is camouflaged under a thick Russian accent.*]

Sasha: Can I join you, friends?

Aviram: Sure, sit down, comrade [*jokingly, offering Sasha a hand*].

Rachel: [*putting her chemistry book on the empty chair, to be sure Sasha doesn't sit there*]: Pull over another chair, Sasha—we're saving this for a friend.

Geula: A friend I think Rochelle has a crush on, as they say in America. He's short, dark, and handsome . . . with that eastern aura.

Sasha: I didn't mean to break up the conversation. And where are all those people leaving the lobby going? [*He gestures to the crowd.*] All those kids with Israeli flags. [*As he gestures to the children, he knocks over the eggs, which splatter on the table and the floor.*]

Aviram: That's O.K. I really preferred them scrambled [*as he wipes with the napkin Geula gives him*]. The crowds, oh, they're headed for the Peace Process demonstration.

Geula: [*bursts in*]: So-called Peace Process. The Labor, Alignment, Peres, Rabin. . . . [*suddenly turns to Aviram*] What did you think of that article, Aviram, is Rabin a *rodef*? It caused somewhat of a stir. [*She remembers to stir her coffee.*]

Aviram: It is dangerous to use such terminology. . . . The implication is that a *rodef* can be killed.

Sasha: What is a *rodef*?

Geula: If it's not an American who quotes Talmud, it's a greenhorn Russian. You don't know anything, do you? [*She winks teasingly but Aviram comes to Sasha's defense.*]

Aviram: Take it easy, Geula, he's new here. You know, Sasha, that the taking of a life is forbidden in the Jewish tradition. But the *halakhah* has some exceptions to this rule—in time of war, for execution after condemnation by the Sanhedrin, and . . . this is where the idea of a *rodef,* the pursuer, comes in. . . . If a man is chasing you, knife in hand, threatening to kill you, it's then allowed to kill him. In self-defense.

Rachel: [*quietly*]: It seems a little far from a man running after you knife in hand to calling the Prime Minister of Israel a *rodef.*

Geula: [*angry*]: You soft woman! There are rabbis in the settlements, better than we, who say it may be so. He [*nods at empty chair*] can lecture for two hours on it. The argument is that the Labor politicians by giving up land are endangering the settlers' lives, and that makes the politicians in effect potential murderers. And secondly, the land they're giving away to the Arabs is Holy Land.

Rachel: [*again softly, but distinctly*]: I think I hear Korah again. . . .

Aviram: No, you women are both wrong. You are like the man in Amichai's poem [*and here he picks up the book and reads*] "who in the middle of a bridge forgets both its ends." The rabbis are hardly Korahites, and Rabin is not a *rodef.* The overwhelming objection to Rabin's policy has nothing to do with messianism—remember most of the Jews who disagree with Rabin are not at all religious. No, the opposition, and I think it is legitimate, comes from a completely different reading of the Arabs' intentions.

You asked about that article? If you read it carefully, the rabbi who wrote it said pretty clearly that Rabin is *not* a *rodef,* because an Israeli government acting in good faith, even if many oppose it, cannot be ruled a *rodef.* So, some rabbis are more radical. . . . They call him a *rodef.* That's just political rhetoric. Even in the Talmud sometimes ideas are stated in hyperbolic language. For example, "Someone who lives outside Israel is as if he is Godless." Or if you don't get the ingredients for the temple incense right, you are subject to the penalty of death. . . .

Rachel: [*takes a napkin, dips it in a glass of water, and wipes some of the egg off of Aviram's sleeve, affectionately.*] And in Tractate Shabbat it says: "A scholar who has a stain on his jacket gets the death penalty!" I'm not kidding.[62]

[*They all burst into good-natured laughter.*]

Aviram: And I thought you were sweet on *him* [*points to empty chair*].

Geula: Aviram is right. When people write articles like that, it isn't only because they feel that when land is given back that the coming of the Messiah is hindered. There is simple fear, that they are responsible for the death of Jews in the future. My brother and his wife and kids are frightened for their lives in Hebron. Ironically, it was Labor who gave them the green light years ago to settle there. And now the government is pulling the carpet out from under. Rabin is *pursuing* peace at a murderous pace. Can you understand that people are desperate?

By giving up Judaea and Samaria, we're guaranteeing, absolutely guaranteeing, the death of Jews. Maybe of your children. The man who authorizes the deal that gives the land back and inevitably leads to the death of Jews—that man is a traitor and a *rodef.* He must be stopped. . . .

Rachel: This is crazy. You say it, and yes, I have heard rabbis say it, without even adding, as I was taught at Stern, *le'fi aniyut daati* (in my humble opinion), or *ilulei de'mistafina* (a phrase of modest hesitation). . . .

Aviram: I think you think too much. [*again, tries to lighten the atmosphere*] This is what comes from teaching women Torah!

Rachel: [*Picks up chemistry tome and mock-throws it at him, as he ducks*]: Wait, let me finish. And as for not giving up an inch of the Greater Israel—look at Solomon making a deal with Hiram, the pagan King of Tyre, in turning over to him twenty cities in Galilee. Not a word in Bible, *midrash,* or Talmud criticizing Solomon!

Aviram: I also hear what Geula is saying. [*Picks up menu and points to Basel Hotel logo.*] She is afraid that what began a century ago in Basel is threatened.

Sasha: Basel?

Rachel: Theodor Herzl—he founded the Jewish State, in a sense, when he convened the First Zionist Congress in Basel, let's see, ninety-eight years ago. Somehow, I don't think Herzl would have wanted us to chase the Arabs from . . .

Geula: [*interrupting Rachel*]: How can you judge when you've lived all your life in the security of . . . New Jersey? You haven't been surrounded by people who hate you!

Aviram: [*pats the rifle slung over the edge of his chair*]: But Geula, we've got an army, and we're strong. If we can't talk to each other, it will be the end of what began at Basel. The tragedy is that both of you [*looks at the girls*] are partially right. Giving back territory is risky; not giving back territory is risky.

The funeral of Yitzhak Rabin. Jewish men are covered by a *tallit* prior to burial. And at State funerals the coffin is draped in a flag. (Photograph by Avi Ohayon.)

Geula: So what are we to do? I'm scared. [*looks again at watch*] Where is Yigal?

Sasha: Wait, listen.

[*At this point the TV that was blaring senselessly in the background is turned up louder. The TV seems to grow, to fill the stage, shrinking the young people, the cafe, the world. All that is left is a screen, and on it plays the videotape made by chance by an amateur of the assassination of Yitzhak Rabin by Yigal Amir. We then see Rabin's coffin, covered with the* tekhelet *and white Israeli flag. [See the illustration on facing page.]]*

Plate 11
A reproduction of the 1885 Rishon Le-Zion flag.

Plate 12
From tallit to flag. Cutout-color prints, New York, 1906.

Plate 13
A photograph of the site, showing at extreme left a piece of Zedek's minibus parked at the scene of the supposed transgression. The correct "No Parking" sign is at right, the ambiguous one between the cars at left.

Plate 14a
Large drops produce the most vivid rainbow colors. Photograph by David K. Lynch and William Livingston.

Plate 14b
U.S. Postage Stamp, "Special Occasions" series, issued October 22, 1988. Stamp design © 1988 USPS. All rights reserved. Can you find a problem in this rainbow?

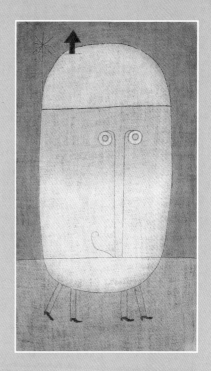

Plate 16
Paul Klee, Maske Furcht (Mask of Fear), *1932. Oil on burlap, 39½" × 22½", The Museum of Modern Art, New York. Photograph © The Museum of Modern Art, New York. © 1997 Artists Rights Society (ARS), New York/ VG Bild-Kunst, Bonn.*

Plate 15
Paul Klee, Eroberer (Conqueror), *1930. Paul Klee Foundation, Museum of Fine Arts, Bern. © 1997 Artists Rights Society (ARS), New York/ VG Bild-Kunst, Bonn.*

Plate 17
Paul Klee, Eros, *1923. Collection A. Rosengart, Lucerne. © 1997 Artists Rights Society (ARS), New York/ VG Bild-Kunst, Bonn.*

Plate 18
Wall mural from the grave of Rekhmire at Thebes, fifteenth century B.C.E., *showing refining (of gold?) and the foot-driven bellows for smelting. Jeremiah was familiar with these processes for purification of metal.*

Plate 19
Yakov Rohamkin, having unpacked his traveling laboratory, searches for shaatnez, *the forbidden mixture of wool and linen, using tools of modern science (microscopes and chemicals that give a color test for linen).*

Plate 20
Nature as a focus for religion: The Ship Rock on the Navajo reservation in New Mexico. Photograph by Kelly West.

Signs and Portents:
No Parking in the Courtroom

A fictionalized version of a real case, tried in an Israeli Court, Beer Sheva, on June 10, 1990. The names have been changed to protect the guilty....

Criminal File 676694/90, parking violation, decided in the Supreme Court of the State of Israel, before Justices Israel, Levi, and Cohen, in the matter of Mr. Ohab Zedek, Appellant, versus the Israel Police, Respondent.

The Appellant alleged that a parking ticket issued by the Israel Police was invalid. The Beer Sheva District Court rejected his plea of not guilty. Zedek appealed to the Israel Supreme Court, which agreed to hear the case because of the far-reaching implications for the issue of public disobedience in the country, the incidence of which has increased in frequency recently.

1. Circumstances of the Case

The alleged crime took place June 1, 1989, at 5:40 P.M. on Ben-Gurion Boulevard, Beer Sheva, Israel. Both Appellant and Respondent agreed that Zedek parked his beige Subaru minibus next to a traffic sign, in front of the Ben-Gurion University Library.

The geography of the site is shown on page 214. Note the ambiguous traffic sign, whose significance is the pivotal issue in this case. A photograph of the site, taken some time after the alleged violation (see Plate 13), was submitted in evidence.

The ambiguous sign began life intended as a standard international "No Parking, No Standing" sign, henceforth to be denoted as "No Standing," illustrated on the bottom of page 214. The ambiguous sign is in the middle, flanked at right by a familiar "No Entrance" sign.

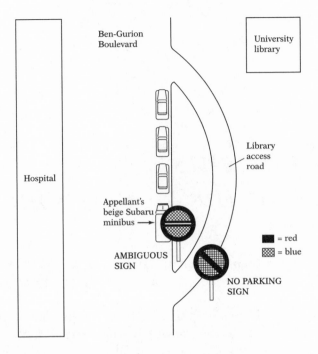

Diagram of the site of the alleged parking violation.

The sign at the scene of the crime was incorrectly positioned, rotated by 45° counter-clockwise (=135° clockwise, given the symmetry of the symbol). The lower court prosecutor elicited in his examination that Zedek's evidence (a sketch (see above) made eleven days

No Standing   Ambiguous sign   No Entrance

The three signs in question.

after the alleged violation, and the photograph (see Plate 13) made three months later) post-dated the violation. The judges viewed this line of reasoning with skepticism, thinking it unlikely that the Israel Police would remount the sign incorrectly *after* a violation.

Zedek claimed that he perceived the sign as ambiguous. He hypothesized it was an Israeli innovation in traffic sign design, based on Jewish tradition, and meant "no entrance to this library access road, except for scholars" or some such message. Jewish tradition is replete with examples of legal restrictions being waived in order to facilitate scholarship[1] and therefore this, he felt, was a reasonable hypothesis.

The Police claimed that Zedek should have realized that it was a misposted, rotated "No Parking, No Standing" sign. Zedek countered that his belief in the infallibility of the Police was so strong that such a hypothesis never entered his mind.

2. Argument of the Appellant

While the facts of the case were reasonably clear, Zedek begged the Court's indulgence in allowing a discussion of the broader issue of signs and symbols, so as to justify his appeal. The Court, after some debate, permitted his unusual multimedia presentation. In doing so it took cognizance of the symbolic nature of the existence of the State of Israel itself, the critical role of signs in the Pentateuch, and the overriding importance of parking to scholars. *

Hockney, Caillebotte, and Maimonides Zedek set the stage with a discussion of an art work by David Hockney. The largest (more than 6 feet × 9 feet) of Hockney's photocollages, it is titled *Pearblossom Hwy. 11–18th April 1986, #2* and depicts a California highway that runs through Mojave, a town northeast of Los Angeles (see the illustration on page 216). David Hockney, born and educated in the United Kingdom, active as an artist there and in the United States, turned in the 1980s to photomontage, the assemblage of photographs, Polaroid or cheaply processed still camera images, into complex collages. Hockney's photocollages are an independent interpretation

---

* A former president of the University of California at Berkeley, Clark Kerr, said "I have come to the conclusion that there are three great problems at Berkeley and they are: sex for the students, athletics for the alumni, and parking for the faculty."

of a cubist perspective—the individual images, sometimes disjointed, follow the movement of the roving eye. They light upon incidental detail (oil cans, beer bottles), conflate space and time (the place to turn, the signs out of proportion). Though Pearblossom Highway seems serenely and totally devoid of human or vehicular motion, Hockney's photomontage makes us move, quickly, on the road.[2]

This collage is symbolic of our century, Zedek noted. Life in California (or Beer Sheva), with its debilitating dependence on the internal combustion engine, is the most extreme manifestation of our addiction to cars and the need for traffic signs to control their use.

Compare the Hockney picture with *Paris Street, A Rainy Day* by Gustave Caillebotte (see page 217). Caillebotte was a young and wealthy member of the Impressionist school. For a long time he was known primarily as a financial supporter and collector of Monet, Renoir, and Pissarro. In recent years his oeuvre has been appreci-

David Hockney, *Pearblossom Hwy. 11–18th April 1986, #2*, Photographic Collage, 198 × 282 cm. © David Hockney. Note plethora of signs.

ated on its own. J. Kirk T. Varnedoe has characterized Caillebotte's vision as shaped "by a strong psychological involvement in his themes and a tensely willful discipline in his work."[3] The strong contention of a complicated perspective, the diverse, independent motions of alienated human beings, the city behind and within their lives, are apparent in this painting.

Paris of the time was a bustling, dirty, and lively city. But traffic signs apparently were as absent a century ago as they are ubiquitous today; none appear in this Parisian scene. They would have been useful. Perhaps they might have saved the life of Pierre Curie, run over, umbrella in hand, by a Parisian horse cab in 1906.

[*The judges grew restive at this point, and the court record includes a discussion among them on the price of tea in China, an*

Gustave Caillebotte, French, 1848–1894, *Paris Street, A Rainy Day,* oil on canvas, 212.2 × 276.2 cm, 1876/77. Charles and Mary F. S. Worcester Collection, 1964.336. Photograph © 1996, The Art Institute of Chicago. All Rights Reserved. Note dearth of signs.

unidentified quote on "The flowers that bloom in the spring, Tra la," and the advisability of allowing appellants to represent themselves. Justice Levi reminded his fellow justices that art is a mirror of life, and that they should forbear and give the Appellant the right to digress, a custom of scholars.]

Zedek continued, pointing out that traffic signs were hardly a modern invention. In order to direct traffic to the biblical cities of refuge (Deut. 19:7) Maimonides rules that signs be posted: ". . . refuge, refuge was written on each crossroad so [those seeking refuge] should know where to turn."[4]

The interpretation of signs depends upon an agreement that a given configuration of geometric elements and text conveys a specific message. Once this configuration is altered, new messages become possible. Zedek turned to biblical commentaries to help him interpret the misposted sign.

Ramban, a Typology of Symbols, and the Rainbow   The medieval Spanish-Jewish philosopher and physician, Ramban or Nahmanides, posits two different modes of symbol interpretation. His discussion is a commentary on the post-diluvian biblical rainbow described in Genesis (9:12):

> . . . My bow I have set in the cloud, and it shall be for a token of a covenant between Me and the earth . . . that the waters shall no more become a flood to destroy all flesh.

Two modern representations of a rainbow[5] are shown on Plate 14.

Type I: Concrete Signs   Ramban offers two ways of interpreting the rainbow. The first is as follows:

> Concerning the meaning of this sign, He [God] has not made the rainbow with its ends bent upward [U-shaped] because it might have appeared that arrows were being shot from heaven, as in the verse, "And He sent out His arrows and scattered them [on the earth]" (Psalms 18:15).[6]

The reference to the Psalms relates arrows shot from a bow to rain pouring from the heavens, a biblical metaphor indicating God's anger. The symbol for the Hebrew month of Kislev, with its heavy rains, is the November/December zodiac sign Sagittarius, a centaur drawing his bow to release an arrow.

Ramban continues to explain why the bow is not U-shaped:

> Instead, He made the rainbow the opposite of this, [with the ends bent downward] in order to show that they are *not* shooting at the earth from the heavens. It is indeed, the way of warriors to invert the instruments of war which they hold in their hands, when calling for peace from their opponents . . . .[7]

This is the first mode that Ramban offers for interpreting signs; the rainbow represents something in our reality, proffered by God as the sign of the covenant He is making. This Type I mode could be called the concrete (or real, representational, and direct) mode.

Signs that fall into this category (e.g., the traffic sign for "school crossing") in general need little explaining, because most humans share the same reality to which the symbol refers. Or so we'd like to think. There is some cultural ambiguity, Zedek admitted, so that a sign for a women's bathroom might not be effective in a society in which women do not wear dresses, or in which men might wear clothing resembling skirts, as in Scotland.

Type II: Abstract or Symbolic Signs Ramban continues addressing the question of why God chose the rainbow to signify the covenant and, not satisfied with the first explanation, offers another one. The second mode of interpretation is abstract, and depends purely upon convention or agreement between people (or between God and people). This will be called the Type II symbolic (or abstract, arbitrary, convention-dependent) mode.

Ramban cites examples of such arbitrary signs:

> And should you want to know how the rainbow can be a sign, the answer is that it has the same meaning as the verse, "This stone-heap be witness, and this pillar be witness" (Gen. 31:52) likewise "For these seven lambs shalt thou take of my hand, that it may be a witness unto me." (Gen. 21:30). Every visible object that is set before two parties to remind them of a matter that they vowed between them is called a "sign" and every agreement is called a "covenant." Similarly, in the case of circumcision, He said, "And

it shall be a token of a covenant between Me and you" (Gen. 17:11) . . . .[8]

Ramban cites first the stone-heap and pillar that sealed a peace agreement between Jacob and his treacherous father-in-law, Laban. The lambs in his next example also signaled a peace pact. Abraham and Abimelech made an oath to end hostilities at a place that was subsequently called, because of this oath, Beer Sheva. The last example is the covenant of circumcision.

According to this Type II mode, the objects in the examples are irrelevant to the content of the covenant; for example, the stone-heap could have been a tree. Similarly, there is no special meaning in the rainbow per se, nor in its colors, shape, or substance; the medium is irrelevant to the message.[9]

We use such abstract Type II symbols frequently. In particular, all signs containing writing are, by definition, symbolic, since alphabets (or word sounds) are arbitrary. Within a literate subculture sharing a language they may be very easily and unambiguously interpreted. Nevertheless, Zedek wondered, what would Ramban make, even if he knew English, of road signs such as "Soft Shoulders Ahead," "Gap in Verge," or "Kiss 'n Ride"?

After outlining this typology, Zedek used it to explain how he, the thinking driver/parker, interpreted the traffic sign in question. A standard "No Standing" sign definitely belongs to the abstract Type II category, because there is nothing in it that relates to the reality of parking or standing (see illustration bottom of page 214). In contrast, the horizontal bar of a "No Entrance" traffic sign puts it into the concrete Type I category, because in human experience a horizontal is used to prevent entrance—old-fashioned door bolts, railway crossing gates. The rotation of the white bar of the sign in question, from a diagonal position to a horizontal one, changed the sign from the abstract type into the concrete type of symbol; it became to Zedek more like a standard "No Entrance" sign (Type I). He did not enter, he parked.

When faced with new or ambiguous signs, every individual must reach some judgment as to the meaning of the sign. A calculus of similarities rather than identities was required, and this may be constructed using the notion of "fuzzy sets."[10]

[*Zedek was about to launch into an exposition of recent advances in fuzzy set theory made by his brother-in-law's cousin, when he noticed signs (Type I, direct) of impatience from several of the judges. So he rested his case.*]

3. Argument of the Respondent

Arguing on behalf of the Police, a sergeant continued the discussion of sign theory that the Appellant had begun, chiding Zedek for his outdated citations. Although Ramban's words were timely and timeless, his last publication was about the year 1260!

Being a dyed-in-the-wool ex-intelligence officer, the sergeant would remove the gray areas of uncertainty surrounding the sign's meaning by using color decoding theory. He cited Umberto Eco's distinction between the private idiosyncratic view of color and the shared public world of color. The linguist author Eco states this with a flourish, "My personal relationship with the colored world is a private affair as much as my sexual activity, and I am not supposed to entertain my readers with my personal reactivity towards the polychromous theater of the world."[11,12] Therefore we must unravel the puzzle of the sign, putting aside our own color predilections. The sergeant continued to echo Eco, "Human societies do not only speak of colors, but also with colors. We communicate with flags, traffic lights, road signs."

The Optical Society of America says humans can distinguish ten million colors! Thus, a traffic control language that would be richly expressive could be composed. But subtlety is not what will stop a Mack truck hurtling down the highway at 70 miles per hour. Therefore, the 116 official Israeli road signs utilize only five colors.

These 116 signs are categorized by color and shape into warning, informative, regulatory, and prohibitory groupings. The disputed "No Standing" sign belongs to the regulatory group with blue backgrounds. By no stretch of the imagination could it be confused by anyone with red prohibitory signs.[13] Of these there are *only two*, because they involve life-and-death situations: "No Entrance" and "Stop."

Colors, too, convey different messages, according to Ramban's scheme. The red in the Ethiopian flag is an abstract Type II representation indicating faith, and in the Benin emblem—soil.[14] But in most flags red expresses the bravery associated with blood and, for the same reason, in traffic signs red means danger. This is a Type I representation: blood$\Rightarrow$red$\Rightarrow$danger. Unless Zedek wants to opt out of his cultural context, he couldn't confound a blue sign that merely regulates the flow of traffic (and is Type II, symbolic) with a red sign (Type I, concrete) which prevents a driver from heading to his bloody death down a one-way street.

The dual coding of signs by color *and* pattern ensures that messages are communicated even if the colors fade because the geometry of the sign remains intact. And vice versa—if the geometry alters as in the disputed sign, the colors remain clear transmitters of police's message not to park. The Police reaffirmed that a transgression of the law had occurred.

4. Justice Israel's Opinion

Justice Israel (see biographical sketch below) opened his decision by noting that Ramban's mode of sign analysis applied as well to representations of reality in the modern science and technology that evoked the sign structure which, in turn, provoked this case. By looking at representation in modern science, the Justice continued, we will be guided to the way in which observant drivers, such as the Appellant (a lapsed scientist, the holder of several degrees, as questioning elicited), should deal with ambiguity.

> *Justice Dan Israel*, born in Padua, first studied chemistry and biology at Hebrew University. His brilliant research on hydrophobic bonding in proteins was, however, so devastatingly criticized by a biochemist that the sensitive young man switched fields to the law. His dislike for adversative rhetoric pushed him toward the role of a judge, and he soon began a rise through the lower courts, culminating in an appointment to the Supreme Court in 1983.

That Jewish legal processes should partake of the reliable knowledge of science is made clear by Maimonides's specification of the qualities of judges:

> Only wise and intelligent men, who are eminent in Torah scholarship and possess extensive knowledge, should be appointed members of either the supreme or lower courts They should be somewhat aware of such fields as medicine, mathematics, astronomy, forecasting constellations, astrology, methods of soothsayers, augurs and wizards as well as idolatrous superstitions, and the like, in order to be competent in dealing with them.[15]

It is essential to the scientific enterprise that physical, chemical, and biological observables be represented, whether it is by a variable in an equation ($F = ma$) or a chemical formula (H_2O), or a picture of a kidney glomerulus. Questions of the reality or faithfulness of such representations ensue: What is mass? Does the water molecule or the glomerulus really look that way? Is it of issue to see what either of these "really" does look like?[16]

Ramban's classification is actually a useful analytical tool, Justice Israel said, in determining the range of ambiguity in various representations in science.

Let us take water as an example. The substance is essential to life, especially to the region where the Appellant lives. It has been the object of human contemplation ever since the events that preceded the rainbow passage explicated by Ramban. And without water there are no rainbows.[17]

There is a problem in representing water because (a) it is a liquid; (b) in small amounts it is colorless and takes on the hue of its container; (c) in large volumes it is colored, due to what scientists now recognize as an "overtone" of a certain molecular vibration, but its color in nature depends on its environment (contrast the North Sea in storm versus the coral atoll of Bora-Bora); (d) we have at hand under normal climatic conditions three phases of this substance—liquid, ice, and steam; and (e) chemists want to know its microscopic molecular structure, H_2O, and how it aggregates in the solid and liquid phases. And that's not easy to see.

The Justice presented a list of various representations of water, to be analyzed according to Ramban's typographies:

1. A photograph of water.
2. An artistic representation of the liquid.
3. Two representations of water in old textbooks.
4. The word for water in various languages.
5. The formula H_2O.
6. A structural formula of the water molecule.
7. A computer-generated dynamic model of liquid water.

1. A photograph is an approximate, two-dimensional representation (of a three-dimensional object) taken with a camera (which

utilizes a lens that may sensibly and in a controlled manner distort the scene) and developed and printed on paper. The light areas in the print are a second-generation (a negative intervened) representation of the areas where light hit the film (dark clumps of silver grains in the negative, absence of those clumps in the positive). As anyone who has made a print from a negative knows, there is a wide range of manipulation of the image available to the photographer. And these considerations do not include modern computerized image modification. Given all this capacity for intervention and manipulation, it's remarkable that the image remains as Type I, direct, as it does.[18]

A little reflection (yes, the light and angle matter) shows that the iconicity of representing this primordial liquid depends more on the container or surroundings than on the water itself. A close-up of a small piece of the photograph below (some water sans surroundings) quickly becomes an abstract, unassignable image.

A photograph of the women's ritual bath (*mikvah*) in Netanya, Israel, for monthly immersion by married women.

2. David Hockney has been drawn to water throughout his career. Representations of swimming pools, in various media, abound in his paintings. *The Splash* (see below) is typical (great) Hockney: wide areas of color, a rather flat representation with minimal perspective, somewhere between abstract and realistic, "cool" (but therefore hot in emotional undertones), a sense of action or movement evoked by an almost cartoon-like schematic essence. There could not be a human figure in this painting, could there?[19]

The Hockney painting is firmly positioned in between Type I and Type II categories. It's "representational" in a minimal way, abstract in many others, beyond the ways in which any painting is not real.

3. On page 226 there are reproductions of two illustrations of water from nineteenth-century chemistry texts.[20]

The general features of such old scientific illustrations are described adequately in the following lines from a lesser contemporary American poet:

David Hockney, *The Splash,* 1966, acrylic on canvas, 183 × 183 cm. © David Hockney.

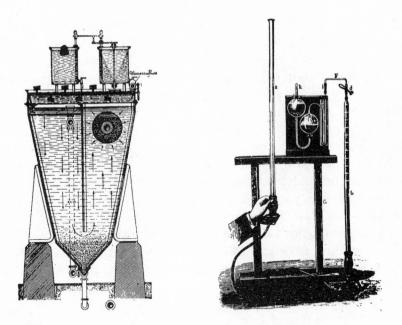

Two nineteenth-century illustrations of water in the laboratory. (*Left*): A water purification system from a German chemistry text, 1900; (*Right*): A gas analysis system from an 1848 French book.

If you look in old chemistry books
you see
all those line cuts
of laboratory experiments
in cross-section.
The sign for water
is a containing line, the meniscus
(which rarely curls up the walls of the beaker),
and below it
a sea
of straight horizontal dashes
carefully unaligned vertically.
Every cork or rubber stopper
is cutaway.
You can see inside
every vessel
without reflections, without getting wet,

and explore every kink
in a copper condenser.
Flames are outlined cypresses
or a tulip at dawn,
and some Klee arrows
help to move gases and liquids the right way.
Sometimes a disembodied hand
holds up a flask.
Sometimes there is an unblinking observer's eye . . . .[21]

This poem goes on to point to a surprising loss in understanding as one moves from abstract to concrete representation. The meniscus and dashed lines in the illustration on page 226 are minimally concrete. It's interesting to speculate, though, to what extent this symbolism derives from the medium (woodcut or engraving) that was used from 1450 to 1900 to reproduce images.[22] The intermittent, broken, dashed line is at the heart of this representation.

4. The word for water is, of course arbitrary. The English expression derives from the Nordic *vatten*, German *wasser*. The word is old, and differs in other languages: *voda* (Russian), *eau* (French), *aqua* (Latin), *shui* (Mandarin). The English letter *M* comes from the ancient Phoenician/Hebrew letter *mem*, written ∿, which symbolized the first syllable of the Hebrew word for water, *mayim*. It's an example of an acrophonic alphabet, where an object is pictured, in this case water as waves, and that picture represents the sound of the first consonant of the word.[23] The waves of water stood for *m* in ancient Hebrew, and in Hebrew script today מ. Eventually this became the English *M*. Most of our letters are derived that way; for example, the Hebrew letter *bet*, from the first sound of the word *bayit*, house, was a boxlike symbol.

Justice Israel now begged the Court to take a jump from the macroscopic world to substances in science, often described in microscopic detail.

5. An ounce of water contains no less than 10^{24} molecules of water, all in mad motion at room temperature. Each molecule is made up of two hydrogen atoms and one oxygen atom, combined in such a way that they stick together. Only at high temperatures (>500° C) would H_2O begin to fly apart into its component atoms.[24]

The symbols H for hydrogen and O for oxygen are just as arbitrary as the word water (or the letters in it). There is, however, nothing arbitrary about the signified atoms nor about the numeral 2 in H_2O, denoting the ratio of the atoms in the molecule. It was only around 1860 that chemists agreed that there were two hydrogens per oxygen and not one. The representation, anyway, is clearly symbolic.

6. The structural formula for water is indicated below. It moves us toward the concrete mode, carrying an implication of a three-dimensional (here actually two-dimensional) structure. Shapes of molecules are critical, determining all their properties. So it's significant not only that water is H_2O, but also that the oxygen is connected to two hydrogens (and not to one, as in H-H-O). And that it is "bent" (H-O-H angle 104.5°) and not linear is critical. Were water linear (H-O-H angle 180°), its properties would be very different and it might not be a liquid at ambient temperatures.

Structural formula of water.

So is this structural formula concrete? No. It's a model, one representation, enlarged about one hundred million times, of a water molecule. Other representations are shown below, one a space-filling model (left), the other an electron density map (right). Each has a claim to the iconic, for each represents, quite nonuniquely, some aspect of the molecule.

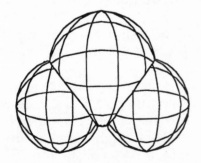
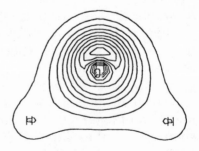

Two modern views of the water molecule. (Drawing by Donald B. Boyd, Eli Lilly and Co.)

7. Can we get a "better" picture of liquid water on a microscopic level; can we approach more nearly the concrete, Type I sign? To some degree.

What one does (using a computer) is to pick a "statistically significant" sample of water molecules, say a hundred of them. Then one sends them on their merry way, traveling in arbitrary directions, with a range of velocities that is set by the temperature. They collide with each other, and with the walls of the containing vessel. This is a simulation, at the molecular level, of real water. A typical picture of the instantaneous positions in this dance is shown below.[25]

Is this "real," putting aside the unreality of the two-dimensional representation? Well, yes and no. It's pretty close to what happens in your seltzer, but still a model.

In concluding his analysis of the range of representations of water, Justice Israel remarked that it was impossible to put the representations numbered 1 through 7 on a single scale of directness/symbolism. Little "runs" of increasing directness might be noted, for example, $5 \rightarrow 6 \rightarrow 7$ or $3 \rightarrow 2 \rightarrow 1$, but a detailed analysis of 7 (the scientific dynamic model) or 1 (the photograph) reveals quickly the unreality of these representations. Perhaps one could mount the representations on a circle, but better still would be an analysis recognizing the ambiguities inherent in each and every representation. And the reality of each as well.[26]

Justice Israel then returned to the case of the Appellant, O. Zedek. While the Justice was sympathetic to Zedek's ambiguity on facing the

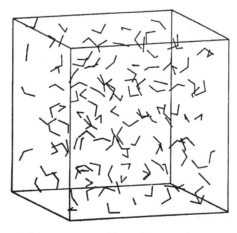

Dynamic simulation of the structure of liquid water. (Drawing by David L. Beveridge and S. Swaminathan.)

mismounted sign, he ruled that the lower court's GUILTY verdict should stand.

He reasoned as follows: A scientist (and he had already mentioned Zedek's advanced degrees) realizes that *any* sign is to some degree arbitrary, as the above analysis of water shows. Furthermore, a scientist is trained as a detective, to recognize anomalies, to reason on the basis of partial knowledge, to mistrust the obvious.

A glance at the faulty sign would certainly show a horizontal bar. But a scientist surely goes beyond the obvious. He notices the bar is not a white one on a red field ("Do Not Enter"), but a white one encased in red on a blue field. At this point the scientist is bound by the dictates of his profession to form alternative hypotheses about the origin of the mysterious sign. Here are some:

1. This is a new sign, whose meaning I don't know.
2. This is a "No Entrance" sign, but someone messed up its painting.
3. This is an incorrectly mounted "No Standing" sign.

Theory, previous knowledge, and new experiment enter into the process of falsifying some of these hypotheses, leaving one standing as the most likely: the "No Standing" one.

As ingenuous as Zedek's statement about his faith in the Police sounds, surely the facts indicate (and here the Justice elicited that Zedek reads his daily newspapers) that the last hypothesis, given Ockham's Razor (that one should not use complex explanations when simple ones suffice),[27] is likely to be correct. It's what Zedek, as a trained scientist should have suspected, and since he didn't, he was GUILTY.

5. Justice Levi's Opinion

The second opinion of the court was delivered by Justice Levi (see biographical sketch below). At the outset he disagreed vehemently with Justice Israel's conclusion. The scientific deliberations recommended by his esteemed colleague would lead to seminars at all four-way stop signs.

Justice Aharon Levi was born in Tel Aviv in 1937. His parents left Israel when he was a child, to seek their livelihood raising chickens in New Jersey. At Princeton University the future justice studied art history, much to the consternation of his parents. On the family's return to Israel he continued his studies, now in law, eventually launching himself into a lucrative private maritime law practice in Haifa. In a brief moment in Israeli politics when complaints arose as to the otherworldly and antibusiness slant of court rulings, he was appointed to the Supreme Court. Justice Levi's private law practice has allowed him to build a fine collection of twentieth-century art.

To Justice Levi the legal question was crystal clear. Drivers have absolute responsibility to obey properly signed traffic codes. And the Police have the responsibility for ensuring that the directives are unambiguously specified. If the signs are incorrectly posted, the driver has no liability. This was upheld in a Supreme Court Appeal (2/3/83, *Isaac Asolin vs. State of Israel*).

The Justice wished, however, to justify his reasoning in still another manner, drawing upon the Appellant's and Respondent's interest in sign structure. Ramban's analysis is clearly related to current theories of linguistics, literary criticism, and anthropology based on the work of Ferdinand de Saussure (1857–1913) and Charles S. Peirce (1839–1914).[28] The Police had assumed the Court's knowledge of these literary matters, but the Justice thought he had better inform his colleagues on such abstruse matters, for which his Princeton education had prepared him.

Saussure analyzed language as one of many systems of signs, and introduced the seminal distinction between the signified (say, the unnatural condition of no cars parked along a street) and the signifier (the physical object directing law-abiding Israeli citizens, as few as they might be, not to park so). Together the signifier and the signified make the sign.

Saussure called the science of signs *semiology*; his American counterpart, Peirce, termed it *semiotics*. Peirce proposed a classification of signs in terms of the relationship between signifier and signified. His system has the feel (and volume) of talmudic discourse.

Peirce identified three functions that a sign might have. It might be an *icon*, by resembling its signified.[29] This is what the Appellant identified as Ramban's Type I or direct mode. And what the Macintosh computer's icon commands have brought to a wider audience. The Type II sign was called by Peirce simply a *symbol,* an arbitrary convention of association. And he distinguished a third function, that of an *index.* The latter operates by some direct connection in space and time, showing a relationship (a pointing finger, a knock on the door).

Now neither Peirce nor Saussure cites Ramban, as a cursory library search by the Justice's law clerk, a lapsed linguist, revealed. Neither do they cite each other. Though we stand on the shoulders of giants, Justice Levi said wistfully, they had better be of the same subculture.[30]

A semiotic analysis of many aspects of our culture is instructive, and traffic signs in particular lend themselves to such discussion. A valuable book by Martin Kampen, *Geschichte der Strassenverkehrszeichen* (*History of Highway Traffic Signs*), is devoted to this subject.[31] It is useful, the Justice continued, to analyze the signs in the Hockney photomontage, presented by the Appellant, classifying them à la Ramban or Peirce, to see what factors might have motivated Zedek to his illegal action in an ambiguous situation.

The "Pearblossom Hwy," "Stop Ahead," "California 138," and "Stop" signs in the Hockney opus (see the illustration on page 216) belong to the symbolic Type II mode because there is nothing in their content or shape that corresponds to the signified object. But what about the right and left arrows under the "California 138" road sign, and the arrow atop the "Stop Ahead" sign (note the symbolic redundancy of this sign—a symbol *and* a text; perhaps Californians suffer from iconic surfeit)? Peirce would classify these arrows as indices. The Ramban would examine each sign in terms of his two categories. It is likely that he would not dichotomize. But to a man living in a world in which arrows were the instrumentality of war, it would be difficult to attach a mere symbolic value to them.

Actually, it is instructive to explore the ambiguity of the arrow sign in the work of an artist arrowmaker par excellence, Paul Klee. Klee's work is shot through with arrows; the symbol appears on the cover of his *Pedagogical Sketchbooks* and in hundreds of his creations. Why the arrow? Because Klee constructed his universe in terms of tensions or balances. One of these was between standing still and moving, between static and dynamic. The arrow was icon as well as symbol of this tension, whether it was physical or within the psyche. Klee also viewed the arrow as an essential symbol of the human condition: "The father of the arrow is the thought: How do I expand my reach. . . ."[32]

In Klee's *Conqueror* (1930; see Plate 15) we see iconic use of the arrow. It might be thought that the arrow is even redundant, for the imbalance of the figure and the standard he bears give this watercolor its motion. But that is not so; if we delete the arrow, the composition's sense becomes one of stumbling rather than conquest. The arrow focuses the other forms in the composition.

A more symbolic use of the arrow is to be found in the oil on burlap painting of 1932, *Mask of Fear* (see Plate 16). This was created at the beginning of Nazi period; Klee's works figured prominently in the notorious 1937 Munich exhibition and sale of "degenerate" art.[33] The upper-left arrow clearly symbolizes constrained thought.

Finally, the arrow can become an abstract symbol, as in the 1923 *Eros* (see Plate 17). The subtly colored stripes gently define the ground of desire where strong forces will join, inevitably, in the realm of ideas or love.

These three works by Klee, a small sample of the arrow-containing subclass of this prolific artist's oeuvre, illustrate the continuous gradation from iconic to symbolic representation, as shown, symbolically, in the illustration below. Or should the arrow run the other way?

In concluding his discussion of arrows, Justice Levi pointed out that the meaning of an arrow in art or in everyday life will depend upon the context. Zedek, faced with the disputed sign, made a reasonable guess about its meaning.

Justice Levi ruled in favor of the Appellant Zedek: NOT GUILTY.

[*At the conclusion of the reading of Justice Levi's opinion, the Appellant interjected that as far as he was concerned, all he saw in*

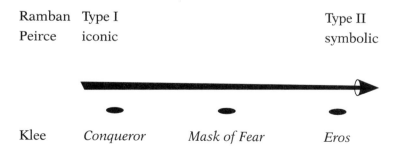

A symbolic classification of some of Klee's works.

*Israel were right turn arrows. The Justices ordered this remark
stricken from the record.*]

6. Justice Cohen's Opinion

Justice Cohen (see biographical sketch below) expressed some dis-
comfort at the artistic and scientific reasoning of the other Justices.
Though in the end he must concur with one of his respected colleagues, he wished to base his decision instead on uniquely Jewish
tradition and law (the *halakhah*).

> *Justice S Cohen* is the scion of two distinguished families.
> His mother is from the Kappah family of Yemenite Jews
> renowned for their translations and commentaries on
> Maimonides. His father is a descendant of Rabbi Samson
> Raphael Hirsch, responsible for the nineteenth-century intel-
> lectual reinvigoration of Frankfurt Jewry. His mother called
> him Saadya, and his father called him Siegfried; to keep the
> peace he adopted the initial S in place of any name at all. He
> describes his parents' "mixed marriage" as "an intercultural
> marriage that blends many traditions." This was a mixed
> blessing: He was expelled from the Hirsch Horev Yeshiva
> because he mixed up the blessings for *matzah* balls and
> gefilte fish, neither of which he recognized, having been
> raised on his mother's Yemenite *kebbaneh*.* For a hobby he
> does ritual calligraphy of biblical verses.

He began by congratulating the Appellant for adducing the arguments of Ramban. The Justice also had some acid remarks on the
lack of corresponding traditional learning among the Police, but
these were stricken from the record. He summarized the problem

---

* *Matzah* balls are soup dumplings made of *matzah* (unleavened bread) meal;
gefilte fish is made of ground up carp, whitefish and pike, spiced and boiled in
water. Both originated in Ashkenazi European Jewry and both are delicious (especially if your mother makes them) small white balls; however they require different
blessings when eaten because one is boiled *matzah* meal and the other boiled fish.
Kebbaneh is a Yemenite specialty bread which requires yet a third type of blessing
since it is baked flour.

before the Court as follows: If the ambiguous sign were to be interpreted in a Ramban Type I, iconic mode, the misposted sign would be invalid. If, however, it were to be read in the Type II, symbolic way, it would be legally valid.

Justice Cohen was heir to both modes. From his father he had received the iconic tradition of sign interpretation, which found its most ardent expositor in the writings of the nineteenth-century German Rabbi Samson Raphael Hirsch. Hirsch emphasizes that a symbol is a piece of reality that conveys to the mental eye a quality difficult to communicate through the mere process of reasoning.

At this point Justice Cohen opened his copy of *Basic Guidelines for a Jewish Symbolism*,[34] a family heirloom printed in Frankfurt in 1858 and dedicated in Hirsch's own hand to the Justice's paternal greatgrandfather. Although Hirsch's work is encyclopedic, there is no direct reference in it to parking signs. But in Hirsch's Bible commentary there is a lengthy discussion of the rainbow, the same sign that occasioned Ramban's semiotic distinction. Hirsch interprets every aspect of the rainbow (shape, substance, color) as a carrier of a message. As the product of the interaction of light and water, for instance, it signifies that

> amidst clouds capable of dispensing either life or death, we behold the presence of light, a reminder that even in the midst of wrath, God's life-preserving mercy endures. . . . The aspect that most approximates the rainbow's symbolic significance is the sign of the spectrum . . . signifying nuances and varieties in the human personality. Now is not the rainbow simply one unified, complete ray of light broken up into seven colors? These colors range from the red ray, which is closest to the light, to the violet, which, farthest from the light (from Heaven), merges into darkness. Yet . . . together, they form one complete white ray. Might this not be interpreted as symbolizing the whole infinite variety of living things from Adam [in Hebrew, "the red one"] closest to God, to the most obscure form of life [represented by violet], a worm?. . . They are all fragments of one life, all refracted emanations of one Divine Spirit.[35]

Justice Cohen said that in the service of fairness he must enter into the record that some critics see Hirsch as reflecting nineteenth-century German romanticism, interpreting Judaism apologetically as an ideal aesthetic system. Reading Hirsch you might think the

Pentateuch was written by Schiller, about whom Hirsch wrote one of his well-known essays.

Nodding to his colleague, he noted there is little arbitrariness in nature, as Justice Israel had shown in the structural representation of a water molecule, where there was even a purpose to the 104.5° angle between the oxygen and hydrogen atoms. So it was unlikely that there would be arbitrariness in the details of biblical symbols.

Analogously, the noticeable rotation of the "No Standing" sign by 45° was certainly enough deviation to invalidate it, given the Hirschean iconic viewpoint.

On the other hand, from his maternal forebears Justice Cohen had inherited the extremely abstract Type II mode of analysis, epitomized in the works of Maimonides, the ultimate rationalist. The Justice then read a segment from a signed copy of the translation by Y. Kappah, his mother's brother, of the *Guide to the Perplexed* (III:26) wherein Maimonides derides those who attempt to find symbolic meaning in every scriptural detail.

> Those who trouble to find a cause for any of these detailed rules, are in my eyes *meshuggah*[36] and devoid of sense; . . . Divine Wisdom demanded it—or, . . . say that circumstances made it necessary—that there should be parts [of His work] which have no certain object. You ask why must a lamb be sacrificed and not a ram? But the same question would be asked, why a ram had been commanded instead of a lamb, so long as one particular kind is required. The same is to be said as to the question why were seven lambs sacrificed and not eight; the same question might have been asked if there were eight, ten, or twenty lambs . . . .[37]

Hirsch, a few hundred years later, goes on in great length about why a lamb, why seven, and so on. The complexity and richness of exegesis lends itself to rococo extravagances in his hands. Hirsch's book was a godsend to Sabbath sermonizers, while Maimonides's exasperation was positioned between the holy—the argument of the Book of Job, that God's will is not for men to fathom—and the human-exaggerated symbolic interpretation.[38]

Justice Cohen, while respectful of tradition, was not opposed to the wisdom of modern science (he brought to Justice Israel's attention the commendable optics of Rabbi Hirsch). Obviously what one had before the Court was a case of pattern recognition.

All signs are to some degree arbitrary; even the most iconic can be deconstructed so that their apparent iconicity is reduced to shambles. Yet people and societies function in spite of the deconstructing mind. The problem of recognizing a rotated sign is similar to identifying rotated or misshapen letters and numbers. Advances in pattern recognition theory have enabled computers to read, for example, zip codes that are imprecisely written.[39]

Justice Cohen cited an example from his own, pre-computer-age heritage. There was a shortage of Hebrew texts in Yemen. So pupils would encircle a teacher who held the sole book available for that group, and they would have to be able to read it from any angle, as we see from the illustration below. Justice Cohen often found himself reading documents upside-down that another Justice, seated opposite him, was holding.

Still, the Justice was ambiguous about the Beer Sheva traffic sign and was of two minds about the guilt/innocence of the Appellant. Therefore, he proposed an empirical pattern recognition criterion, modeled on that for Jewish ritual inscriptions. The test is designed to ascertain how much a handwritten letter deviates from the ideal standard form for that letter. Serious deviation would render the entire document (for instance, a Torah scroll) ritually invalid. The differences between Hebrew letters are often subtle, equivalent to the difference, say, between the English capital letters

Yemenite children reading the same text from different angles.

O and D. Precision is imperative. The test is described in the codes of law. The *Shulhan Arukh* (Tefillin, 32:16) says "If one finds a letter that is unclear, and a child who is neither overly bright nor overly dull can read it, it is kosher (valid). Otherwise, it is invalid."[40] Justice Cohen said that just the previous week a friend brought a document to him to ascertain whether a given letter was ritually valid. In such a situation he calls in his six-year-old son and asks him to read the letter, when the surrounding letters are covered over. If this young, barely literate "unbiased bystander" can read the ambiguous letter, it is deemed valid.*

At that point, the Justice asked the clerk to open the door, and in came the shy lad. Justice Cohen showed him a chart with the ambiguous traffic sign on top, and three other signs on the bottom ("No Entrance," "No Parking," and a correctly aligned "No Standing" sign). He asked the boy which of the three bottom signs most closely resembled the disputed sign. Smiling at what seemed to be a game, the child pointed without hesitation to the "No Standing" sign.

Frowning, Zedek realized that it was almost unnecessary for Justice Cohen to announce his decision: the sign was valid, the Appellant GUILTY.

7. Verdict and Sentence

In conclusion the Supreme Court held, by a two to one vote, that the District Court's judgment was upheld, and Ohab Zedek was GUILTY. However, this being Israel, and there being precedent, the Court ruled as follows. Considering that the fine for the offense of Ohab Zedek was originally 42 New Israeli Sheqels (NIS; $21 at that time, 1990), the Court ruled that the fine would be reduced by half. Thus they sentenced Zedek to pay 20 NIS ($10) or, alternatively, spend one day in jail.[41] Furthermore, the Police of the State of Israel were ordered to pay the expenses of the Appellant.[42]

---

* The Justice noted the relationship of this ancient procedure to the current U.S. test for obscenity, established in 1973 by the U.S. Supreme Court in *Miller vs. California:* One of the three criteria a work must meet if it is to be judged obscene is that "the average person, applying contemporary community standards" would find that the work taken as a whole appeals to prurient interest.

Pure/Impure

> *I have made you an assayer of My people*
> *—A refiner—*
> *You are to note and assay their ways.*
> *They are bronze and iron*
> *They are all stubbornly defiant;*
> *They deal basely*
> *All of them act corruptly.*
> *The bellows puff;*
> *The lead is consumed by fire.*
> *Yet the smelter smelts to no purpose—*
> *The dross is not separated out.*
> *They are called "rejected silver,"*
> *For the Lord has rejected them.[1]*

Jeremiah 6:27–30

Shira Leibowitz Schmidt: In this jeremiad, the prophet berates his people for having gone astray. His language is strong, high, and poetic. And it is interspersed with several passages that indicate substantial familiarity with metallurgy.

An interpretation has been provided by the much-maligned former American president, Herbert C. Hoover, who was a talented, unusually well-educated mining engineer, and by his wife, Lou H. Hoover.[2,3,4] The Hoovers discern in the Jeremiah passages the ancient process of cupellation: An impure mixture of silver or gold with undesired admixtures is melted in a cupel, a shallow dish shaped from bone ash. Lead is added. A blast of air oxidizes the nonprecious metals. The base metal oxides dissolve in the lead oxide, which is skimmed off, leaving behind the pure silver or gold. Plate 18 shows some Egyptian metallurgical practice contemporary with the Prophets. Jeremiah invokes the process metaphorically; the wickedness of his people is so great, they will not be purified. The Hoovers remark:

From the number of his metaphors in metallurgical terms we may well conclude that Jeremiah was of considerable metallurgical experience, which may account for his critical tenor of mind.[5]

Jeremiah's stern criticism caught my eye in its conjoining of a scientific or technological argument and an appeal to purity, a condemnation of mixture. Purity is a traditional feature, indeed a desired goal, of religious systems.

Roald Hoffmann: That passage from Jeremiah is not as clear as you think, Shira. If toward the end the prophet uses a powerful metallurgical metaphor for purity, he undermines his aim by invoking iron and bronze near the beginning. Jeremiah's assessment of these metals as "stubbornly defiant" (emphatically repetitive in Hebrew, *sorerei sorerim*) admits their strength as materials.[6]

And why are they strong? *Because* they are impure, mixtures, alloys. I suspect that Jeremiah, good metallurgist that he was, knew that bronze, in the swords and ploughshares of the Israelites, was a mixture of copper and another metal, tin. And carbon in iron strengthens it; properly processed it becomes steel.

Science teaches us that nothing is pure; moreover that complete admixture is the natural course of events. And chemistry gives us abundant examples of superior impure materials.

So religion squares off against science once again,[7] purity versus impurity. Or so it seems. . . .

I. Religion and the Aspiration to Purity

Colette: As that word "pure" fell from her lips, I heard the trembling of the plaintive "u," the icy limpidity of the "r," and the sound aroused nothing in me but the need to hear again its unique resonance, its echo of a drop that trickles out, breaks off, and falls somewhere with a plash. The word "pure" has never revealed an intelligible meaning to me. I can only use the word to quench an optical thirst for purity in the transparencies that evoke it—in bubbles, in a volume of water, and in the imaginary latitudes entrenched, beyond reach, at the very center of a dense crystal.

The Pure and the Impure[8]

Roald Hoffmann: Words such as "pure" and "impure" carry a multitude of meanings. The sense least encumbered with moral connotation is that describing the distinction between objects composed of one substance versus those consisting of a mixture of several. So Vicks® Throat Lozenges are a mixture of benzocaine, cetylpyridinium chloride, menthol, camphor, eucalyptus oil, D&C Red No. 27, D&C Red No. 30, flavor, polyethylene glycol, sodium citrate, sucrose, and talc. Of the two ingredients labeled by D&C (Drugs and Cosmetics) numbers, D&C Red No. 27 is tetrabromotetrachlorofluorescein; D&C Red No. 30 is 6-chloro-2-(6-chloro-4-methyl-3-oxobenzo[b]thien-2 (3H)-ylidene)-4-methyl-benzo[b]thiophen-3(2H)-one, alias "helindone pink CN" (if that helps. . . .)[9] "Flavor" certainly contains several molecular components. Other examples are your breakfast cereal—read the ingredients! (see the illustration on page 242)—and pure mountain spring water (*certain* to contain, at the parts per million level, calcium, magnesium, chloride, sodium, sulfate, bicarbonate, and organic matter, and at the parts per billion level, all kinds of things you don't want to know about, such as ammonia, borate, fluoride, iron, nitrate, potassium, strontium, aluminum, arsenic, barium, bromide, copper, lead, lithium, manganese, phosphate, and zinc).[10]

From that reasonably neutral starting point of mixture the meaning of "pure" and "impure" develops. First, there is a metaphorical movement to the realm of the emotions, carrying with it a certain confusion with the ideas of concentration and intensity. A saint who meditates, be he Buddhist or Christian, is pure in soul. He is not distracted, he is intense.[11]

Second, the movement to the figurative sphere inevitably triggers the association of a positive ethical or moral value to the pure, and a negative one to the impure. To be spotless, unblemished, to be pure in mind, is to approach godliness. "How blest are those whose hearts are pure: for they shall see God," Jesus says in the Sermon on the Mount,[12] echoing the Twenty-fourth Psalm:

Who may ascend the mountain of the Lord? Who may stand in His holy place?—He who has clean hands and a pure heart, who has not taken a false oath by My life or sworn deceitfully.

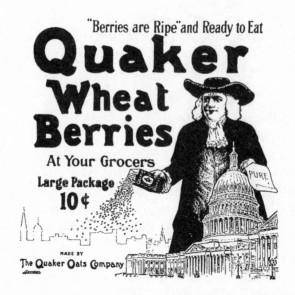

A 1910 advertisement for Quaker Wheat Berries touting their purity. And invoking patriotism in the process. On a contemporary Instant Quaker Oatmeal box you'd find a list of ingredients including rolled oats (with oat bran), calcium carbonate, salt, guar gum, caramel flavor, reduced iron, vitamin A palmitate, niacinamide, pyridoxine hydrochloride, thiamine mononitrate, riboflavin, and folic acid.

The Buddha says in his last sermon:[13]

All composite things are by nature impermanent. Work out your salvation with diligence.

To be pure is to testify to the holiness of God and His people. Purity becomes symbolic, good and *of* God.[14]

Shira Leibowitz Schmidt: Is there any doubt that purity is a positive good of religion? It is an important factor behind the complex rituals and regulations governing marriage, inheritance, sacrifice, and cooking in Jewish observance. Entire tractates of the Talmud are devoted to the rules and regulations of ritual and physical purity. To an outsider the discussion might seem esoteric, a debate between rival rabbinical schools about how many drops of milk accidentally spilled into a veal stew will cause the dish to become thereby a forbidden milk/meat mixture.[15] But for the Jewish people, every act must be a sanctification of His holiness. The exhortation to purity is there in the Torah, the Five Books of Moses:

You shall be holy, for I, the Lord thy God, am holy. . . . You shall observe My laws. You shall not let your cattle mate with a different kind; you shall not sow your fields with two kinds of seed; you shall not put on cloth from a mixture of two kinds of material.

Leviticus 19:2, 19

Rationalist attempts to find hygienic or scientific arguments for these rules, or to seek their economic origin, abound. In *shaatnez,* the prohibition of mixing wool and linen (see Plate 19), some people see the ancient struggle between shepherds and farmers.[16] But while these explanations are ingenious, tracing the inevitable inter-relatedness of the spiritual and physical world, the reasons for the ubiquity of proscriptive ritual must be deeper.

Mary Douglas: . . . as we examine pollution beliefs we find that the kind of contacts which are thought dangerous also carry a symbolic load. This is a more interesting level at which pollution ideas relate to social life. I believe that some pollutions are used as analogies for expressing a general view of the social order . . . ideas about separating, purifying, demarcating, and punishing transgressions have as their main function to impose system on inherently untidy experience. It is only by exaggerating the difference be-tween within and without, above and below, male and female, with and against, that a semblance or order is created.

Purity and Danger [17,18]

Roald Hoffmann: In her perceptive 1966 book, *Purity and Dan-ger,* Mary Douglas views rites avoiding pollution or impurity as rit-ual demarcations. Douglas notes that what attracts the attention of the Lele people of the Congo region about the scaly anteater is that it is *as an animal* abnormal—it looks like a lizard, but unlike a lizard it is a mammal and suckles its young; it is scaly like a fish, but climbs trees; its young are born singly, or twinned in pairs, as those of humans. She constructs what seems to me a plausible parallel argument for a cultural basis of the prohibited animal species of the Jews, the so-called abominations of Leviticus.[19]

Douglas goes on to argue persuasively that "where the social system requires people to hold dangerously ambiguous roles, these persons are credited with uncontrolled, unconscious, dangerous, disapproved powers—such as witchcraft and evil eye."[20] The disordered, or that simply outside the ordered, is not just static and expelled. It spells danger to a stable society. Danger is power.

For a stable society, or a stable form of matter, danger may be as simple as change. In a "phase transformation," which is the precipitous change of one form of matter to another (for instance water to steam or ice), the beginning of change is always at a locus of disorder, or an impurity.

Shira Leibowitz Schmidt: In her book, Douglas correctly describes the immense importance of separation in the biblical scheme, and its relationship to purity and holiness. This idea is articulated in the blessing said at the end of the Jewish Sabbath, setting it off from the weekdays:[21]

You have graced us with intelligence. . . . You have distinguished between the sacred and the secular, between light and darkness, between Israel and the peoples, between the seventh day and the six days of labor. . . . Our Father, Our King, begin for us the days approaching us . . . free from all sin, cleansed from all iniquity.

But an anthropologist analyzing the terminology of purity in a foreign culture through the veil of English will be handicapped. First of all, highly developed religious systems do make a definite distinction between physical and ritual impurity.[22] Thus in Hebrew there is *tahor*—pure, clean physically (and spiritually, by metaphoric extension). This can be negated, as *lo tahor*—impure, not clean physically and spiritually. But spiritual defilement, pollution, is described by another adjective, *tamei*. The biblical candelabra was made of pure (*tahor*) gold, but it might or might not be *tamei*, depending on its contact with a source of defilement (a corpse, a reptile, etc.).[23]

Even if anthropologists study a culture in terms of its own language, they may be comparing—really confusing—concepts that seem similar but aren't. There are biblical laws that seem superficially to be about the same concerns: mixtures and impurities. These laws are related to gastronomy (meat/milk, leaven/non-leaven), fabric (linen/wool), animal breeding (horse/donkey), and marriage (Israelites/neighboring pagan peoples), as well as time (*eruv tavshilin*, mixing of holidays and Sabbaths), place (*eruv tehu-*

mim, mixing of public and private domains), and metaphysics (impurity due to proximity to sources of ritual defilement). But each concept is embedded in its own legal infrastructure from which it cannot be extracted and compared with others out of their contexts.[24]

This problem is endemic to anthropological approaches. In a deconstruction of the marriage laws in India, Wendy Doniger recently observed that "the attempt to rationalize other people's apparent irrationalities is a game that many scholars of religion have enjoyed playing, particularly . . . in this era of moral relativism."[25]

Ultimately, the strong claims of religions do not depend on what men and women call reason. This has been the conclusion of Jewish thought—witness the Book of Job, or Rav, a third-century C.E. sage who, commenting on a passage from Psalms (18:31), "the Lord's utterance is pure," asserts "what difference does it make to the Holy One whether one eats unclean or clean substances? It follows that the commandments were given only to purify people."[26]

George Frideric Handel: [27]

A part of the score of the Chorus, "And He Shall Purify," of Section 7 of Handel's *Messiah.*

Roald Hoffmann: The connection between purity and religious aspiration is awesome. Listen to this oratorio in which George Frideric Handel set several lines of the prophet Malachi's call to the music of thirty singers and an equal number of instruments:

> *And who shall stand when he appeareth?*
> *For he is like a refiner's fire,*
> *And like fuller's soap;*
> *And he shall sit as a refiner and purifier of silver;*
> *And he shall purify the sons of Levi,*
> *And purge them as gold and silver.*

Malachi 3:2–3

Jeremiah was not the only prophet who knew metallurgy.

II. Science and the Drive Toward Impurity

Nicolás Guillén:

> *I am not going to tell you that I am a pure man.*
> *Among other things*
> *we have yet to know if what is pure exists.*
> *Or if it is, say, necessary.*
> *Or possible.*
> *Or if it tastes good.*
> *Have you ever had chemically pure water,*
> *laboratory water,*
> *without a grain of dirt or excrement,*
> *without a bird's small excrement,*
> *water composed only of oxygen and hydrogen?*
> *Puah! What filth!*
>
> *I do not say, then, that I am a pure man,*
> *I will not tell you that: everything to the contrary.*
> *That I love (women, naturally,*
> *for my love can speak its name),*
> *and like to eat pork with potatoes,*
> *and chickpeas and sausages, and*
> *eggs, chicken, lamb, turkey,*

fish and clams;
and I drink rum and beer and brandy and wine,
and fornicate (even on a full stomach).
I am impure, what can I say?
Absolutely impure. . . .

From "I Declare Myself an Impure Man," 1973[28]

Roald Hoffmann: When a chemist runs any reaction in the laboratory or sees some immunosuppressant activity in an extract from a fungus, the product at hand is almost certainly a mixture. Guillén's "water composed only of oxygen and hydrogen" indeed not only tastes flat to us but is unnatural.

Why all that impurity? In the realm of the living, that's an easy question to answer. A living organism is complex. Even within one specialized cell there are thousands of chemical reactions going on. The multitude of tasks accomplished is staggering—can I begin to describe what transpires as the energy of light, and water, and carbon dioxide combine in a chain of reactions in which we have identified dozens of steps (and a role for iron, copper, manganese, and magnesium) on the way to a sugar molecule in a lily-of-the-valley?[29] More than the hundred or so elements, it's the millions of molecules that we sculpt from them that shape the chemistry of the universe. The complexity of plants, our own complexity, demands variety: Two hundred seventy-five constituents have been identified in rose oil, a particular rose essence extract.[30]

So mixtures and impurity are natural. More than that—there is a natural drive to maximize mixing, called entropy.[31] It's not a matter of actively seeking the most messy state of the universe. It's simply the democratic principle of all possible states being equally likely. Even with small numbers the tendency to disorder is evident. If you toss a coin three times you will get one of the equally likely results HHH, HHT, HTH, THH, TTH, THT, HTT, TTT (H = heads, T = tails). The two "pure" results (HHH, TTT) are unlikely, *not* because on a given toss there is a preference for a single head or a single tail (they're equally likely for the fair coin assumed), but because such a result represents only one of eight possible equally likely outcomes. Imagine for 10^{23} coins (that's how many molecules there are in a slurp of water) how improbable it is that all 10^{23} should give heads on a toss! As improbable as that a toss of a properly mixed Caesar salad should lead to three kinds of lettuce neatly segregated, the anchovies on top, the cheese underneath, the egg reconstituted, croutons all together.[32]

It's not far from an identification of the pure with the good to the impoverishing notion that behind every observable of this world is a single cause.[33] In biology, as Thomas Eisner has mentioned to us,[34] such ideas have positively inhibited understanding. Take the one gene–one enzyme theory, or the assumption that each insect sex attractant must be a single molecule.

Insects have a chemically rich communication system. Sometimes a male of a species can detect a handful of molecules of one kind wafted by a female among billions of extraneous molecules. Such "pheromones" have been sought and isolated, a triumph of modern chemistry.

The story of the sex pheromone of the cabbage looper moth, *Trichoplusia ni,* is instructive. The pheromone was first (1966) thought to be a single molecule [(Z)-7-dodecenyl acetate]. Then in 1980 a second crucial component was identified, and in 1984 a clue in the way the main component was synthesized by the insect caused Wendell L. Roelofs and his colleagues to seek other components, finally demonstrating that a total of no less than six different molecules was involved.[35] The western pine bark beetle, an economic pest, has an aggregation pheromone, a mixture that signals all comers, male or female, of that species, to assemble. It is a blend of three molecules: one from the male, one from the female, and one, remarkably enough, from the tree.[36] These are not isolated examples; most pheromones are blends.

The entropy of the universe increases. We may be able to reverse that trend locally, grow a nearly perfect crystal, write a poem, bring a child to term. But this can be done only with an input of energy, at a cost. A price that makes some other part of the universe messier.

Robert Louis Stevenson: (from Henry Jekyll's full statement) My provision of the salt, which had never been renewed since the date of the first experiment, began to run low. I sent out for a fresh supply, and mixed the draught; the ebullition followed, and the first change of colour, not the second; I drank it, and it was without efficiency. You will learn from Poole how I have had London ransacked; it was in vain; and I am now persuaded that my first supply was impure, and that it was that unknown impurity which lent efficacy to the draught.

The Strange Case of Dr. Jekyll and Mr. Hyde [37]

Roald Hoffmann: Not only is there a natural tendency to mix, but chemists and physicists are constantly made aware of the occasionally superior properties of imperfect, disordered matter (glass is such), or of composites. Or the pheromones just mentioned. Or the valued outcome of the human perfumer's art. This was certainly appreciated by the metallurgist side of Jeremiah, familiar with bronze and brass, knowing solders and precious metal alloys. Bronze, an alloy of copper and tin (or other elements; the first bronzes were alloys with arsenic[38]), has mechanical properties superior to either pure metal (see the illustration below[39]). It was common in the weapons, tools, and decorative objects of the biblical

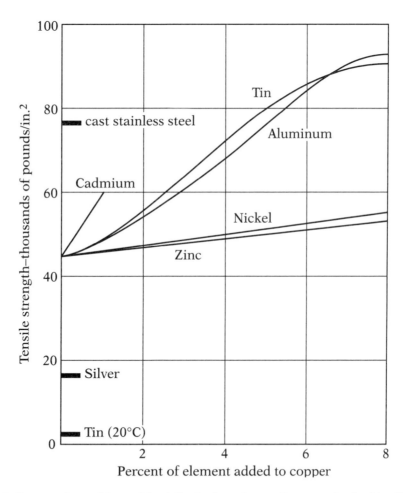

The influence of metal impurities (alloying) on the tensile strength of cold-rolled copper. The bars at left indicate the tensile strength of tin, silver, and one kind of steel. Note that bronze (copper alloyed with tin) is stronger than either pure copper or pure tin, and may be as strong as steel.

period. Pure metals, and even more so alloys, are strong and ductile precisely because of the existence of imperfections (called "dislocations") in their structure.[40]

A Sumerian disputation: Silver, only in the palace do you find a station, that's the place to which you are assigned. If there were no palace, you would have no station, gone would be your dwelling place. . . . In the (ordinary) home, you are buried away in its darkest spots, its graves, its "places of escape" (from this world). When irrigation time comes, you don't supply man with the stubble-loosening copper mattock, that's why nobody pays any attention to you! When planting time comes, you don't supply man with the plough-fashioning copper adz; that's why nobody pays any attention to you! When winter comes, you don't supply man with the firewood-cutting copper ax; that's why nobody pays any attention to you! . . . Silver, if there were no palace, you would have neither station nor dwelling place; only the grave, the "place of escape," would be your station.

Copper's speech to Silver[41]

Shira Leibowitz Schmidt: The above text (the illustration on page 251 shows a transcription of the original tablet[42]) dates to about 2000 B.C.E. It is a "debate" between silver and copper. It may be that copper here really stands for a copper alloy, arsenical copper, or bronze, more likely to have been used in tools than native copper by this time. [43]

This debate is not unique in Sumerian literature. The Sumerians were a combative, verbally aggressive people. We have evidence of this in their texts, their litigiousness, and their legal codes. The level of verbal invective and confrontation in their surviving writings is remarkably high.[44]

Cyril Stanley Smith: No metallic material has had more influence upon man's history than iron and its simple alloy with carbon, steel. . . . Steel differs in composition from pure iron essentially only by the presence of a small amount of carbon. . . . The relation between properties and compositions was fairly clear in the case of the bronzes. . . . The fact that steel was also an alloy was not so clear; indeed it was not definitely accepted until the very end of the eighteenth century [of our era!], 3000 years after the practical discovery. This knowledge arose out of and contributed to the Chemical Revolution in an intimate way.

The Discovery of Carbon in Steel [45]

Roald Hoffmann: An instructive story of the resistance of science to the evidence of the senses (and their extension, instruments), is to be found in the history of the establishment of the

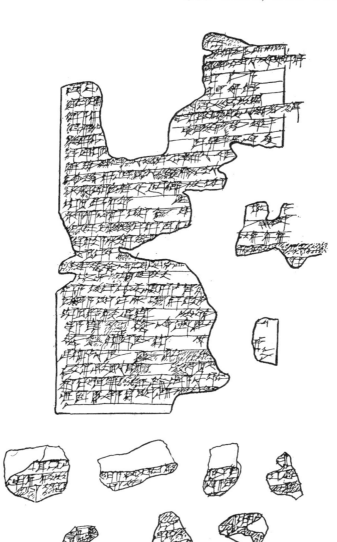

A transcription of one of the cuneiform tablets of the disputation between copper and silver.

correct composition of steel. The material is not new—think of medieval Japanese swords or Damascus steel. Steel is an alloy, but— and this was a large part of the difficulty metallurgists faced in thinking about its structure—not an alloy of a metal, iron, with another metal. Steel is an alloy, an intimate mixture on the atomic level, with a *non*metal, carbon. And the carbon sneaked in, so to speak, through the carbonaceous fuel used in the inevitably intimate contact of heating. Moreover, the optimum admixture of carbon into iron is small, no greater than 1.5 percent, so it was difficult to detect.

Cyril Stanley Smith, a metallurgist who was very much interested in the interface of science and art, tells beautifully the story of the establishment of steel as an iron–carbon alloy. In the early part of Smith's story the scientists and philosophers don't come off too well:

At the end of the seventeenth century, then, we have the practical man (guided as he always will be by the knowledge in his fingers and his eyes) unconsciously putting carbon into iron by his steelmaking processes, while the philosopher thought that some deleterious principle was being removed.[46]

The story unfolds contemporaneously with the ascendance and passing of an erroneous but plausible theory of chemical reactivity, phlogiston.[47] Experimentation and the phlogiston idea (despite its being wrong) convinced people—correctly—that something was being added to iron to make steel. Next, practical observation and careful analysis led Swedish chemists to conclude that what was present besides iron was a carbonaceous residue called "plumbago." Interestingly, the earliest written record of carbon in steel is in John Pettus's *Volatiles from the History of Adam and Eve* (1674) which mentions "charcoal" unconsumed by fire rising out of molten cast iron and uses this as a metaphor to bolster man's hope for resurrection.[48] In France, in the years just before the Revolution, within the framework of a revolutionary theory of chemistry, the admixture in steel was identified as carbon.

Mircea Eliade: It has been established that among miners, rites calling for a state of cleanliness, fasting, meditation, prayers and acts of worship were strictly observed. All these things were ordained by the very nature of the operation to be conducted because the area to be entered is sacred and inviolable; subterranean life and the spirits reigning there are about to be disturbed; contact is to be made with something sacred which has no part in the usual religious sphere—a sacredness more profound and more dangerous. There is the feeling of venturing into a domain which by rights does not belong to man—the subterranean world with its mysteries of mineral gestation which has been slowly taking its course in the bowels of the Earth-Mother.

The Forge and the Crucible [49]

Roald Hoffmann: There is an anthropological and religious dimension to the art of winning metals from their ores and alloying them, as evidenced by the elaborate rituals surrounding primitive mining and metallurgy.

Extending to inanimate ores and metals the life-giving sexuality of the biological world made eminent sense. The idea was especially strong in China, given that civilization's philosophical acceptance of the *yang* and *yin* cosmological principles and its metalurgical skills. The "marriage of metals" that is alloying is an ancient notion, reflected, as Eliade points out, in the *coniunctio* or *Chymical Wedding* of alchemy (the illustration below is one of the less explicit representations of this union). And perhaps in another guise in Hegelian dialectic, or the dualities that pervade the world.[50]

Scientists continue to make happy marriages of the elements. In the 1980s a new class of materials was synthesized, ceramic in

A view of the *Chymical Wedding,* from the *Viridarium Chymicum . . .* (Chymisches Lustgärtlein) of Daniel Stoltzius von Stoltzenberg, 1624.

nature and conducting electricity without resistance at temperatures much, much higher than other previously known superconductors. The first superconductors, discovered early in this century, were pure metals. The world record holders until recent times were alloys of two elements. The new superconductors are compounds of four, five, or even six elements.[51]

The illustration below shows a structure of one of these superconductors. Not only are there several elements in it, but for the ceramic to exhibit its properties a certain departure of the oxygen concentration from an integral atom ratio, a certain measure of nonstoichiometry, as it is called, is needed. So the superconductor is not only a mixture, but also "imperfect," departing from the simple composition our naive minds want it to have.

Superconductors, superalloys. Conferring power, if not danger. Or maybe danger, too, steel in the unpredictable hands of man. It seems that nature is following Mary Douglas's vision of the impure imparting power.

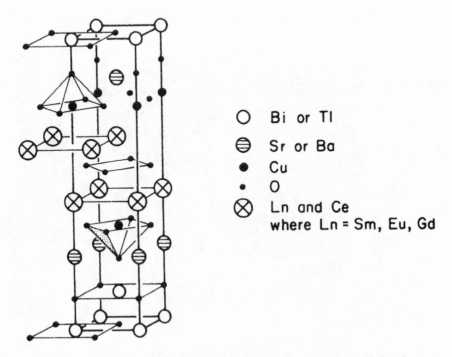

O	Bi or Tl
⊖	Sr or Ba
●	Cu
·	O
⊗	Ln and Ce where Ln = Sm, Eu, Gd

The molecular structure of one of the new superconductors. Note all the elements it contains.

Shira Leibowitz Schmidt: Given all this you might imagine that scientists should come out clearly on the side of mixing, disorder, and impurity; whereas the stance of religion, to which we already alluded, would be the opposite, advocating a pure soul and body. But things are not so simple.

III.   There Is Also Aspiration to Purity in Science

Roald Hoffmann: The extraordinary properties of some substances emerge only when they are pure. Examples are polymers, such as polyethylene (used in food wrap and myriad other products), which are stronger when they are crystalline and pure. Silicon, in computer chips and transistors, has to be made exceedingly pure before it exhibits its immensely useful semiconducting properties. But then, to confound things, the pure silicon is "doped" by "impurity atoms" (roughly one impurity to every 10^7 atoms of silicon). This enhances the electronic efficiency of the purified silicon by many orders of magnitude.

Our favorite metal, copper, whose strength was increased by admixture of tin (or aluminum, or nickel; see the illustration on page 249) has its prized electrical conductivity *decreased* by the very same alloying. The purer the copper, the better it serves us, electrically.

Thomas Morton: Underlying everything we do is the notion of a pure compound—without that concept we'd be powerless (which is one of the obstacles in studying olfaction, since subjectively "pure" odors are often complex mixtures, while chemically pure compounds often possess subjectively "complex" odors). The definition of purity allows us to function with purpose.[52]

Roald Hoffmann: What Morton, a thoughtful organic chemist, says has both historical and philosophical import. Chemistry developed by refining methods of separation, isolation, and purification. If we need to worry about harmful dioxins at the parts per billion level, we must have methods of detecting and identifying them in a complex mixture.

Philosophically, it is impossible to define impurity in the sense of mixture, without the notion of purity or the unmixed. Only the establishment of one substance as a pure compound allows one to call another a mixture.

Recall the *blend* of chemicals that characterizes a typical natural insect communication system, the cabbage looper moth sex pheromone mentioned above. Roelofs and co-workers identified six compounds, in definite proportions, as being necessary. How did they do so? By synthesizing all six separately, then concocting blends that would elicit male flights against an opposing airstream in a wind tunnel.[53]

In science, especially in chemistry, we are continuously engaged in the dialectic of purity, and are made aware of the complementary claims of the simple and the mixed. Purity is a Platonic ideal. Never attained, but quintessential in defining our aspirations. . . .

Aldous Huxley: To make a tragedy the artist must isolate a single element out of the totality of human experience and use that exclusively as his material. Tragedy is something that is separated out from the Whole Truth, distilled from it . . . chemically pure. . . . It is because of its chemical purity that tragedy so effectively performs its function of catharsis.[54]

IV. Mixtures and Impurity in Religion

Preliminary morning service: The incense was composed of the following eleven kinds of spices: balm, onycha, galbanum, and frankincense, seventy *maneh* weight of each; myrrh, cassia, spikenard, and saffron, sixteen *maneh* weight of each; twelve *maneh* of costus; three *maneh* of an aromatic bark; and nine *maneh* of cinnamon. [Added to the spices were] nine *kab* of Karsina lye, three *seah* and three *kab* of Cyprus wine—if Cyprus wine could not be obtained, strong white wine might be substituted for it—a fourth of a *kab* of Sodom salt, and a minute quantity of *ma'aleh ashan* [a smoke-producing ingredient]. Rabbi Nathan says: A minute quantity of Jordan amber was also required. If one added honey to the mixture, he rendered the incense unfit for sacred use, and if he left out any of its ingredients, he was subject to the penalty of death.[55]

Shira Leibowitz Schmidt: While this admonition to good practice upon penalty of capital punishment should be posted in all chemistry laboratories, the recipe given clearly describes the blending of a ritual prescription, of a mixture. The quoted passage from the morning service, originally in the Talmud, is striking counterevidence to the seemingly inevitable emphasis on purity in religious ritual.[56, 57]

Nothing can be omitted, not even the foul-smelling galbanum. Its admixture serves the Talmud as a morality metaphor:

A prayer quorum in which none of the sinners of Israel participates is no prayer quorum; for behold the odor of galbanum is unpleasant, yet it was included among the species for incense.[58]

Another example of required mixtures: In normal clothing, the observant Jew should avoid the forbidden mixture of wool and linen, *shaatnez*, as we mentioned earlier. But in Exodus 28, in the detailed prescriptions for building the Tabernacle and the attendant rites, the robes of Aaron and his sons, the priests, are to be woven with gold, violet, blue, and scarlet wool yarn, and fine linen, all five materials twisted in each thread.

The fact that linen/wool *shaatnez* mixtures are, according to the circumstances, sometimes forbidden and sometimes required, suggest that there isn't anything intrinsically bad—or good—about such combinations. Maimonides puts this philosophically when he places these precepts under the rubric of the command to control one's desires:

A man should conquer his passions, and is warned not to say, "By nature, I do not lust after prohibited things. . . . I am repulsed by meat mixed with milk, I am repulsed by *shaatnez*, I am repulsed by forbidden sexual unions." Say instead, "They are attractive but what can I do? My Father in Heaven has forbidden them!"[59]

In its penchant for specificity, the Talmud deals repeatedly with the problem of defining loss of purity by mixture with the impure. Depending on the sanctity of the commandment, either a majority rule applied, or the prohibited object had to be nullified by a large excess, 60 to 1, of the allowed or pure. In some cases, as for leavening in Passover, no nullification was possible.[60]

Roald Hoffmann: There is an interesting analogy here to the Delaney Clause, a controversial amendment of the U.S. Federal Food, Drug and Cosmetic Act, which for nearly two decades in principle banned the addition to food of all substances that were shown to be carcinogenic, at *any* level, in humans or test animals.[61]

Shira Leibowitz Schmidt: In evaluating the permissibility of admixture, rabbinical law made a substantive distinction between intended and fortuitous acts. There's a wonderful account of this logic in the resolution by Rabbi Tobias Geffen of Atlanta of a problem facing the observant Jew wishing to imbibe Coca-Cola.[62] Rabbi Geffen inquires in 1935 of the secretive and successful Coca-Cola company about their ingredients, a guarded formula known only to a few officials. He discovers that one ingredient (named only "M" to respect the penchant for confidentiality of the soft drink company) is made from meat and fat tallow of nonkosher animals, another

("A") is made from grain kernels, kosher but constituting leaven, therefore impermissible at Passover.

Ingredient M is there only in one part per thousand (as Geffen, properly cautious, has the chief chemist of the state of Georgia establish). No matter that that is below the usually permitted *halakhic* admixture rate of 1:60 in Jewish law—M nevertheless makes the beverage unacceptable, because its inclusion is intended, an act of volition.

Rabbi Geffen does not give up:

Because Coca-Cola has already been accepted by the general public in this country and in Canada, and because it has become an insurmountable problem to induce the great majority of Jews to refrain from partaking this drink, I have tried earnestly to find a method of permitting its usage. With the help of God, I have been able to uncover a pragmatic solution. . . .

He learns from some chemists that M can be replaced by a plant oil such as cottonseed oil, and A by an extract of sugar cane. Astonishingly, he convinces the Coca-Cola company to make this substitution! Since this is America, you can predict what follows (see the illustration on page 259).

Roald Hoffmann: I recently had occasion to use this new-found knowledge of the religious distinction between intentional and accidental admixture. After a poetry reading, I was chatting with an older woman. When I told her that we were working on a book on science and Jewish tradition, she proceeded to tell me an anguished story.

This woman's son had married an observant Jewish woman, and the young couple were keeping a kosher household. One time, when the mother was visiting and minding the cooking, a neglected veal stew was about to burn. In desperation, the mother seized the nearest spoon to stir the pan. Her daughter-in-law, upon entering the kitchen, noticed that the stirring spoon was from the set of dishes reserved for dairy food. She proceeded to give her mother-in-law hell for such a mistake, which caused the daughter-in-law to dump the whole stew, and expend much effort in elaborate cleaning of the pot and spoon.

The older woman was clearly distraught. I was able to calm her down, and assure her (though I'm hardly a rabbi), that she did nothing wrong. The stew was fine. The "contamination" clearly was minor, below the 1:60 ratio, and ever so clearly accidental. I tried to stay out of the real reasons for the conflict.[63]

The certification (in English and Yiddish) that Coca-Cola is kosher for Passover.

Mary Douglas: The final paradox of the search for purity is that it is an attempt to force experience into logical categories of non-contradiction. But experience is not amenable and those who make the attempt find themselves led into contradiction.

Purity and Danger[64]

Roald Hoffmann: Douglas reaches for an existentialist conclusion to the question why, if what a culture chooses to see as dirt and pollution is judged repulsive, do so many of the rites use the unclean, impure, even abominable substances. The Lele, mentioned before, have a cult of the scaly anteater, the creature that seems to contradict the usual animal categories. The attendant rituals properly performed, the animal is eaten. It brings fertility to Lele women and success to its hunters.

Douglas thinks that the incorporation into human rite of absolutely everything, even the "unnatural," is inevitable. It is nature's way and so the way of societies.

It is thus evident that neither the realm of religion nor that of science is unequivocal in its stance on purity. The sacred texts of this world, despite their aspirations to purity, acknowledge the claims of inevitable, even desired admixture. And science, which first recognized the inherent tendency of the universe to mix, nevertheless strives mightily for the purity that is indispensable for a definition of impurity.[65]

V. Metaphors That Compel

Tacitus: For myself, I accept the view that the peoples of Germany have never contaminated themselves by intermarriage with foreigners but remain of pure blood, distinct and unlike any nation. . . . Silver and gold have been denied them—whether as a sign of divine favour or of divine wrath, I cannot say. . . . Even iron is not plentiful; this has been inferred from the sort of weapons they have.

Germania[66]

Roald Hoffmann: Advocacy of purity seems harmless. More than that, as an inducement for the betterment of a human being in body and soul, for throwing off an idol or a drug, it serves us well. The quest for purity is normative, it describes how we *should* be. However, what gives me pause about the positive validation of purity is its abiding use by most known nationalist totalitarian movements, from fascism to apartheid.

The appeal sounded by such regimes and their propagandists is beguiling. It goes as follows: Once upon a time the people were pure in body and mind, beautiful and strong. Then they were corrupted by outside forces, by foreign gods. If only the people returned to this natural pure state, if they expelled the foreign elements, if they cast out the admixture in spirit and flesh of the outsider, then, once again, the people, ah the people, would be fair and strong.

Never mind the diverse ethnic origins of Italians—such shining phrases were at the heart of Mussolini's fascist appeal. Never mind that Turks, Jews, Ukrainians, and Russians have lived among Moldavians for centuries—that's the battle cry of today's Moldavian nationalists, passing laws to keep all those others out of "their" university in Kishinev. And who marches in the vanguard of these racist, divisive forces—the intelligentsia, the religious, and the young people!

It seems clear to me that demagogues of every ilk have intuited what Mary Douglas has told us (and that scientists know, even if they don't want to think of it in this way): Impurities are dangerous—they hold power, the power for change.

Shira Leibowitz Schmidt: I beg to differ, Roald. One must separate the legitimate from the nefarious in ethnic pride. The desire of a nation, which may be small in number, to feel it must aggressively guard its language, customs, culinary habits, and religious traditions if it is to maintain its ethnic identity (in an age when Coca-Cola is the least common denominator for all cultures) is understandable.

In Jewish law, anyone, absolutely anyone, who is willing to adopt the regimen of true observance can join the tribe. The archetypical example is Ruth, the Moabitess, mother of Davidic royalty. But a small people cannot extend the same toleration to those of its own people who want to go the other way and leave the tradition. A chemical analogy to the sweet and strong bounds of tradition might be a semiporous osmotic membrane. A barrier that permits one-way flow in, but not out.

Michel Tournier: Purity is the malign inversion of innocence. Innocence is love of being, smiling acceptance of both celestial and earthly sustenance, ignorance of the infernal antithesis between purity and impurity. Satan has turned this spontaneous and as it were native saintliness into a caricature that resembles him and is the converse of the original. Purity is horror of life, hatred of man, morbid passion for the void. A chemically pure body has undergone barbaric treatment in order to arrive at that state, which is absolutely against nature. A man hagridden by the demon of purity sows ruin and death around him. Religious purification, political purges, preservation of racial purity—there are numerous variations on this atrocious theme, but all issue with monotonous regularity in countless crimes whose favorite instrument is fire, symbol of purity and symbol of hell.

The Ogre[67]

Roald Hoffmann: It may be that the scary similarity of religious and totalitarian appeals to purity derives from their parallel rhetorical structure. The aim is to convince, with words. The situation before Mussolini or Jeremiah is the following, human one: "We are in (physical, emotional) state **I** (for impure), which I do not like. I want to exhort you to move to state **P** (for pure)." The exhortation naturally takes the form of postulating a prior state **P′** and the presence of a disturbing factor **X**.

But rhetoric is not ethics, which is what got rhetoric into trouble. There is a world of difference, an ethical and spiritual essence of a difference, between fascist (or ethnic Moldavian) claims and Jeremiah. Which the parallel rhetoric, parallel guiding metaphors, disguise.

Primo Levi: The course notes contained a detail which at first reading had escaped me, namely, that the so tender and delicate zinc, so yielding to acid which gulps it down in a single mouthful, behaves, however, in a very different fashion when it is very pure: then it obstinately resists the attack. One could draw from this two conflicting philosophical conclusions: the praise of *purity*, which protects from evil like a coat of mail; the praise of *impurity*, which gives rise to changes, in other words, to life. I discarded the first, disgustingly moralistic, and I lingered to consider the second, which I found more congenial. In order for the wheel to turn, for life to be lived, impurities are needed. . . . Dissension, diversity, the grain of salt and mustard are needed: Fascism does not want them, forbids them, and that's why you're not a Fascist; it wants everybody to be the same, and you are not.

The Periodic Table[68]

Roald Hoffmann and Shira Leibowitz Schmidt: The pure/impure dichotomy is another aspect of the central theme of man and the universe. Its other incarnations are the one and the many, the same and not the same, natural and unnatural. If there be more than one of any thing, and a way of naming or classifying each, if there be a choice, material or spiritual, the problem of purity will arise. A motion in one direction, say to segregate, inevitably raises the opposite possibility.

Two extreme arguments could be made. First, there is the line we might today call "entropic," that the natural is the most mixed. So the true path need be that of miscegenation. Support for this way could be adduced from hybrid vigor in biology. Another, contrary philosophy departs from the basic fact that the human body, in its intense local order, is inherently "contraentropic." And so we are fated to support in our thought systems and societies an opposition to mixing. To yield to disorder would be tantamount to surrendering our place in the scheme of things.

We find validity in both views, and no inconsistency in a philosophy that encompasses the two. Only change is eternal. The segregated, isolated, pure, and the completely mixed impure are each in its own way motionless and impotent. Everything else is tense, poised between pure and impure, ambiguous, therefore interesting. Alive.[69]

Camel Caravans in the Pentagon

Roald Hoffmann: This is a story of books, a story that will move, as Jews have moved, from a small town in the Austro-Hungarian province of Galicia to Offenbach, outside Frankfurt in Germany, to the Pentagon in Washington, D.C., to the battlefield in Korea in 1952, to Meknes, Morocco, to Netanya, Israel.

It is not a sad story, but it begins in the direst of times. In 1942, I was Roald Safran, a five-year-old boy in the Galician town of Zloczow (pronounced ZWO-tschov) in what is now Ukraine, but then was Poland. The population of the town was about evenly divided between Poles, Ukrainians, and Jews.

On June 22, 1941, the German army had attacked Russia, marching quickly through the Russian-occupied part of Poland where we were. On July 3, 1941, a Ukrainian-led pogrom drove 2,500 Jews to their deaths at the Jagellonian citadel above Zloczow, most at the hands of a unit of Nazi S.S. Einsatzsgruppe C. My grandfather Wolf was killed there, my uncle Abraham, with a dum-dum bullet in his wrist, saved himself by playing dead, and later crawling out from the mass grave.

The remaining 8,000 Jews of Zloczow were confined to a ghetto. In three German "actions" in 1942 and 1943 the ghetto was liquidated, its inhabitants sent to Auschwitz, Belzec, and Sobibor.

We had been moved earlier, my parents and I, to a labor camp, a source of slave workers toiling to fix the roads and bridges the war had quickly ruined. Some of these structures my father, Hilel Safran, a civil engineer, had built before the war. Eventually, in January 1943, my mother and I were smuggled out of that labor camp, and were hidden for the remainder of the war, some fifteen months, by a friendly Ukrainian.

Zloczow's Jewish community was in transition, even before the war. There are photographs of my greatgrandparents, clearly pious Jews. My grandparents were observant, my parents socialists, training in the summer for founding those Israeli collective farms that Geula Abramovitz disparaged (see the last scene of Chapter 5, "The Flag that Came out of the Blue"). The town had been a center of *Hasidism* in its time—it was the home of Yehiel Mikhal, the Maggid of Zloczow, a legendary figure in early *hasidic* history. And from Zloczow came (to New York City, as I eventually did) one of the greatest of Yiddish poets, Moshe Leib Halpern.

Books and paper were valued in peace; they were treasured in war. The same small notebook in which my father wrote in Polish chapter summaries of the relativity theory text that he was reading (in the labor camp!), was the only place my mother could write

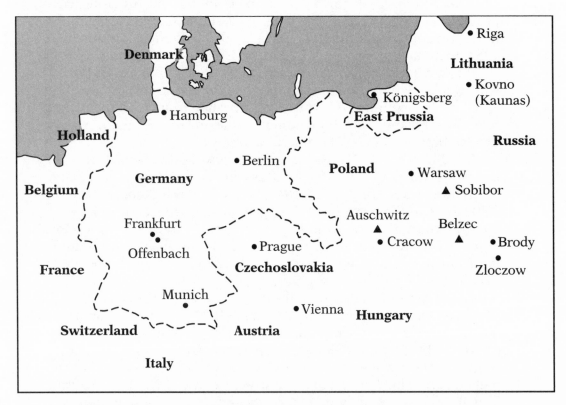

Map showing the location of several of the towns and countries mentioned in this chapter. The border marked is that of pre–World War II Germany. At some time during the war, German rule extended to every town on this map.

down her anguish when she learned of my father's death in a betrayed breakout attempt he led.

The religious Jews of Zloczow brought to the ghetto their books. As the ghetto was liquidated, books were taken from the synagogues, houses of study, and private homes by the Nazis and their helpers. Prayer books, Torah scrolls, Talmud volumes, venerated collections of responsa, *midrash* sets, commentaries, in Yiddish and Hebrew—were hauled off.

The books were neatly crated . . . but then people also were told nice stories about where the cattle cars were taking them. The people died, but as it turned out, some of the books were spared.

Alfred Rosenberg, the chief Nazi ideologist, rose to fame by writing several books of racist theories and anti-Semitism. He had been appointed by Hitler in 1940 to head the Hohe Schule, the future ideological University of Nazism. For this "educational" purpose Rosenberg's emissaries ransacked Jewish libraries all over Europe, and concentrated the gleanings in Offenbach (see map on facing page and photograph below), outside of Frankfurt (ironically one of the

S.A. (Sturmabteilung) officers in Hamburg confiscating books, May 15, 1933. (Photograph by Joseph Schorer.)

earliest centers of Jewish printing, flourishing since 1512, and one of the greatest). There Rosenberg amassed the treasure trove of Jewish books he would need for his "university." He thought his students should hold in their own hands the work of the Jewish devil.[1,2]

Back in Galicia, the few of us who survived (perhaps 150 out of 8,000 in Zloczow, and I, one of three children in the group) were freed by the Red Army in June 1944. My mother Clara and I began the long journey that five years later would take us to New York. Along the way she remarried another survivor and we all acquired the name Hoffmann; but that is another story.

West of us, the war still raged. In 1945, when the Germans surrendered, the great General George Patton, "Old Blood and Guts," he with the pearl-handled revolver, found himself in possession of the Offenbach books, perhaps "the greatest collection of Hebrew books and manuscripts that mankind ever saw or ever will see."[3] He didn't know what to do with this wondrous hoard; Hebrew books were not Patton's forte at that time.

Eventually it was decided to disperse the Offenbach collection to several institutions. The lion's share went to The Hebrew University in Jerusalem. One of the smaller institutions that received several of the Offenbach crates was the Jewish chaplaincy of the U.S. Armed Forces. Some of the Offenbach books were to prove helpful in solving problems that arose for Jewish soldiers in the U.S. Army in subsequent decades.

Shira Leibowitz Schmidt: In 1950, a year after Roald came to the United States, the Korean War broke out. There is a long history of the military providing religious counsel to its soldiers—the illustration on page 267 shows an 1870 kerchief by an anonymous artist, commemorating the Day of Atonement service of 1,200 German-Jewish soldiers outside the besieged French city of Metz.[4] Note the focus on the ark containing the scroll, the Torah being read in the field.[5]

As the Korean War heated up, the Pentagon initiated a policy of moving its chaplains up to the front with the troops. They asked the leadership of each religious group to specify the minimal religious paraphernalia that their chaplains would need to fit into their already heavy kitbags. They asked Cardinal Spellman's Military Ordinariate and the reply was that the Catholics could make do by dedicating any little stone and placing it on a table as an altar. They asked the Protestants, and their needs could be met by providing small Bibles and hymnals. They asked the Jewish Welfare Board

Torah scroll in an ark, at a Day of Atonement (Yom Kippur) service for German-Jewish soldiers outside of Metz, 1870. Commemorative kerchief. © The Israel Museum, Jerusalem.

Chaplaincy Commission whether their chaplains could manage with a Torah scroll that the Pentagon was willing to photograph and reduce to portable proportions. To answer the Pentagon, the Jewish Chaplaincy Committee appointed a Legal Committee composed of rabbis which began to study the relevant sources.

To understand the complexity of the problem, we must know something of the making and use of a Torah scroll. From earliest times, the Five Books of Moses were inscribed by specially trained scribes on parchment, animal hides soaked in limewater, stretched, and scraped. The more than three hundred thousand letters of the

Torah must be handwritten on skins by a trained scribe using a feather or reed pen dipped in specially prepared ink. The skins are then sewn together with kosher animal sinews, and rolled on two strong wooden rollers. The Torah scroll is encased in wood or covered with cloth, and often decorated with silks and silver appurtenances.

The reading of the Torah scroll from parchment is a central event of Jewish public worship,[6] and only then are the blessings before and after the reading said. There are other differences between a parchment and paper Pentateuch. One that is printed on paper and bound into a book must be handled with care and respect; a handwritten Pentateuch on parchment wound in Torah scroll form requires much more stringent care. One example: A Torah scroll must never be turned upside down, something that might be hard to prevent when it is transported to, from, and around a battlefront. When a Torah has to be shipped somewhere, to prevent its ever being in such an upside-down position, one of the seams holding the parchment skins together is undone, so that there are two incomplete scrolls, neither of which has the level of sanctity of a whole one.[7]

In order for the Legal Committee to answer the Pentagon's query about using a miniature, photographed Torah scroll at the front, the rabbis had to study the responsa that dealt with related questions over the centuries. Many of the sources they needed turned out to be among the books that had been rescued by Patton's troops at Offenbach and sent to the Chaplaincy Committee. Two of these rabbinic decisions are of particular interest.

Roald Hoffmann: One responsum was written by a rabbi I feel particularly close to, Zvi Hirsch Hayyot (pronounced in our Galician dialect as CHA-yes). He had been asked a related question about printed Bibles. Rabbi Hayyot was born in 1805 in Brody, a Galician town 20 miles north of my Zloczow (see the map on page 264). Two centuries later, I feel a proximity to him in spirit as well as geography.

Hayyot's father, a Florentine banker, had settled in Brody. Near that town, my father worked on his first assignment as an engineer more than a century later. As a boy, Hayyot studied Italian, German, and French with his father, and Latin, history, and science with tutors. He became expert in the Talmud and commentaries at a young age, and subsequently innovated the use of modern critical methods in talmudic studies. One of Hayyot's works is available in English.[8] He corresponded with leaders of enlightenment Jewry in Galicia and

Italy, and when Austrian law required rabbis to take the university examinations, he was the only traditional rabbi who voluntarily complied (and consequently earned a doctorate). In Rabbi Hayyot's works we see the mind of a modern Orthodox scholar, one of the first to study the nature and authority of tradition. Some of this is apparent in what he wrote in response to questions related to printing.

One of his predecessors, Rabbi Ezekiel Katzenellenbogen in Hamburg (b. 1714), had reasoned that printing could not be used for ritual purposes because, if God had so desired, He would have revealed the technique to Moses or Solomon, rather than to Gutenberg. Rabbi Hayyot demurs:

> With all due respect to Rabbi Ezekiel, that is no way to argue. Of course, there has been no greater prophet than Moses, and no wiser man than Solomon . . . but God, who foresees the future, has arranged for progress of the ages and that each generation should make its discoveries. . . . There is great benefit that will accrue from the printing of Torah, the ability to disseminate Torah in thousands of copies. In addition, printing greatly benefits the world at large.[9]

In fact, in the centuries following the supplanting of parchment by paper, and of scribal writing by printing, that consensus had already emerged.[10] Printing could be used for prayerbooks, for marriage documents, for Bibles used for study and nonceremonial use, for Talmuds and commentaries. These would have some level of sanctity—thus requiring that when worn out or damaged they be disposed of in special *genizah* collections, and given a respectful burial. But neither printing nor paper could be used for the inscriptions placed in a ritual doorpost (*mezuzah*), in the phylacteries worn during morning prayer, or for a Torah scroll used in synagogue reading.

Incidentally, such a *genizah*, a storage place for sacred books awaiting burial, was discovered in Egypt in 1896. In time, some 90,000 manuscripts (not necessarily from parchment) were uncovered in the Cairo Genizah, the richest, oldest source of Jewish texts from the late talmudic period (500 C.E.) The caves at Qumran that held the Dead Sea Scrolls may have also served as *genizot*. [See the illustration on page 270.[11]]

Shira Leibowitz Schmidt: You are always digressing, Roald. Back at the ranch, I mean the Pentagon, one needed those guidelines for Jewish chaplains in the battlefield. The Jewish Welfare

(*Top*): A segment on leather from the Dead Sea Scrolls, first century C.E. © The Israel Museum, Jerusalem; (*Bottom*): from the Cairo Genizah, a letter on paper by Judah Halevi, twelfth-century C.E., the great Hebrew poet and scholar of Al Andalus. The Jewish Theological Seminary of America, New York.

Board had appointed a Legal Committee. Just as we saw in the politics of those students around a café table at the Basel Hotel, Jews divide in many ways in their religious observance. Rabbi Solomon B. Freehof, who was a member of that committee, has summarized the problem succinctly: "I was to give an answer that had to be acceptable to the Orthodox rabbinate and bearable to us." Freehof, a great bibliophile, from whose autobiographical remarks we draw much of our material, was affiliated with the Reform movement.

Rabbi Freehof found a precedent for making a ruling in an unusual twentieth-century responsum written by Rabbi Joseph Messas (1892–1974; born in Meknes, Morocco; served in Tlemcen, Algeria, and Meknes). A layman who had attended Sabbath services while at a Mediterranean resort told Rabbi Messas that a cantor had read there from a printed Pentateuch, not from the Torah scroll. The cantor had called people up to the reading stand, and blessings had been recited before and after each *aliyah*. Was this process correct according to Jewish law?

Rabbi Messas replied that such action involved pronouncing blessings gratuitously, a serious violation in that it could be interpreted as possibly taking God's name in vain. The rabbi vehemently rebuked the cantor, sending him a lengthy poem (!) detailing the violations and exhorting him to cease this practice. To support his conclusion Rabbi Messas writes:[12]

> I have a copy of a responsum of the rabbis of Marrakesh, who were asked about a group of more than ten Jews who embarked on a camel caravan across the Sahara. . . .

Evidently, they had room only for a printed Pentateuch; a heavy Torah scroll from parchment would have been the straw that broke the camel's back (see the illustration on page 272[13]).

> The Marrakesh rabbis replied that to have a public reading and pronounce the blessings is certainly forbidden. . . . A public reading from the printed Pentateuch without a blessing for a week or two is also not permitted (although each individual should read the weekly portion at some point). But if people will be traveling for many weeks . . . then they may hold readings but omit the blessings.

The Legal Committee (ruling on the Pentagon's proposal) stood on the shoulders of Rabbi Messas of Meknes (who ruled on the problem at the seaside resort) who stood on the shoulders of the

Two details from Bartolommeo di Giovanni's *The Story of Joseph,* one depicting Joseph being sold to a camel caravan. Fitzwilliam Museum, Cambridge.

rabbis of Marrakesh (who ruled on the problem of the camel caravan). The Legal Committee in the United States concluded:

> Since these photographed *Sefer Torahs* [Torah scrolls] are exactly like the written *Sefer Torahs,* except that they are not on parchment and sewn with sinews, but do resemble them more than [printing] they may be used in Service in the emergency of field conditions, but no blessings should be recited over the reading. Of course, in the camp-chapels, regular *Sefer Torahs* should be used.[14,15]

Roald Hoffmann: From camels to caissons. . . . But Shira, we have gone too far without learning some of the chemistry and materials science of the writing surface and ink.

Torah scrolls must be written on the skins of kosher animals (sheep, cows, goats, deer, gazelles). Vellum, leather, and parchment are all terms for appropriately processed animal skins, quite different from plant-origin paper.[16]

Vellum (literally, veal skin) is a special fine, thin, soft material made specifically of calfskin, and often from uterine calves, whereas leather and parchment can be made from any of the above animal skins. Even if leather and parchment come from the same animal, they will have differed in their method of preparation and in their final form. Leather, the older process of the two, is animal skin treated with tannins coming from oak bark or gall nuts (swellings covering insect eggs on trees). The Dead Sea Scrolls are written mainly on leather.

About the fifth century parchment came into general use; by the ninth century we begin to find Jewish texts written on parchment. An animal skin is soaked in limewater (a CaO solution) to dehair it. Another ingredient used for dehairing in the past (and by a contemporary English parchment maker!) was dog dung, whose bacteria and pancreatic juices assisted in breaking down skin fibers and fat. The skin is then stretched on a frame or kept under restraint to prevent crinkling.

Animal skins and plants, the sources of those materials on which humans have been writing, are all natural materials of biological provenance. That skins are inherently "biodegradable" you can tell by looking at (and smelling) an animal carcass over a period of days. One purpose of the chemical treatment that transforms these natural materials into leather or paper is assuredly to *arrest* that biodegradability, an interesting thought in our age of concern with the immense scale and persistence in the environment of some of our creations. We are most grateful that the Cairo Genizah scrolls biodegraded only a little; we are not grateful if our ten billion styrofoam cups degrade too slowly.

Our skin, the skin of animals, is a wondrous material of many uses. (A schematic cross-section of animal skin is shown on page 274.) The dermis contains the strong molecular structures that give leather and parchment their special properties of strength *and* flexibility. One of these is a protein, collagen, organized into fibers, and not easily digested.

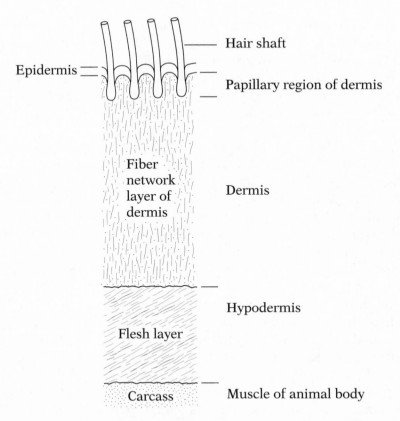

A schematic cross-section of animal skin; the outer layers on both sides are removed chemically and mechanically in the making of leather, vellum, or parchment.

The epidermis carries hair; if you're making a fur coat, it goes without saying that it's important to keep the hair. In the making of leather and parchment, however, the epidermis, and of course the easily decomposing flesh layer, are removed, mechanically and chemically. *Tawing* is the fine English word for this process.

To say that leather is made by tanning hides is only . . . skin deep. The word derives from one important source of the key group of chemicals in the process—the crushed bark of an oak tree (called *tan*). Many vegetable sources can produce the active "tannins," as these chemical compounds are called. The mixture and the process have hardly been simple; here is a list of the ingredients used for tanning in Egypt and Mesopotamia about 1500 B.C.E.:[17]

> Powdered alum for tawing; oak galls, pomegranate rind, sumac berries, cypress leaves, sunt pods (of *Acacia arabica* and *A. nilotica*)

for vegetable tannage; olive oil, animal fats, eggs, milk, myrrh (a resinous plant gum) for oil tannage.

The hides are scraped, then placed in a complex chemical mixture, the exact recipe varying from place to place. The epidermis is partially digested, the hair loosens. Some of the liquid penetrates the dermis, which swells. The tannins begin to do their work.

Shira Leibowitz Schmidt: Forgive me for interrupting, Roald. That tanning process can be quite smelly; the stench from a tannery is like that from a slaughterhouse. For that reason in the Talmud there is a ruling that tanneries must be kept 50 cubits (about 28 meters or 92 feet) downwind from a town.[18]

But if beauty is in the eyes of the beholder, then fragrance may be in the nostrils of the sniffer. And there is something that can overcome the fetid odor of a tannery: the *Shekhinah*, the presence of God. In an olfactory parable, the mystical book of *Zohar* explains how God will accompany Israel even after He scatters them in exile:

> *A parable:* A man loved a woman. She lived in the tanners' market. If she were not there, he would never enter the place. Since she is there, it appears to him as a market of spice-peddlers with all the world's finest aromas in the air.[19]

The parable interprets Leviticus 26:44: "Yet, even then, when they are in the land of their enemies [the tanners' market] I will not reject them or spurn them." The Zohar then asks "Why?" and answers "Because of their bride."[20] The bride is the *Shekhinah*. The parable continues:

> For I love Her! She is the beloved of my soul dwelling there! [The tanners' market] seems filled with all the finest aromas of the world because of the bride in their midst.

Roald Hoffmann: You're always citing passages that show Jewish tradition as fair, democratic, and good. Somewhere else it says:

> The world cannot exist without a perfume-maker or a tanner. Happy is he whose craft is that of a perfume-maker and woe unto him who is a tanner by trade.
> *Our Rabbi taught:* He whose business is with women has a bad character, e.g., goldsmiths, carders, hand-mill cleaners, peddlers,

wool-dressers, barbers, launderers, blood-letters, bath-attendants and tanners. Of these neither a King nor a High Priest may be appointed. What is the reason? Not because they are unfit, but because their profession is mean.[21]

It's lucky there were no chemists then, or we'd get put in the same category, I'm sure.

But there is some splendid chemistry in tanning (or the lime-water processing in the making of parchment) that I have to get in. How do the tannins work on a molecular level? They get at a substance called collagen in the animal pelt dermis.

Collagen is the most abundant protein of vertebrates. It is a beautiful helically braided bundle (see the illustration below[22]) of three polypeptide helices (polypeptides are chains of amino acids, making up the protein of wool as well). The individual polypeptides form left-handed helices; their braid is right-handed.[23] Remember, it's important to tell left from right!

The triple helix of collagen. Drawing and © by Irving Geis.

Collagen is strong, but tanning makes the chaotic assemblage of collagen fibers in the dermis stronger still. The reactive tannins (produced in the fermentation of oak bark and other organic matter) are a class of molecules called polyphenols. They enter this tangled microscopic forest of the epidermis, and bind collagen fibers to one another. In the jargon of the trade, they cross-link the biopolymer. (A similar "cross-linking reaction" takes place in the making of synthetic rubber.) The leather becomes both stronger and resistant to bacterial attack. But it remains flexible. This is why animal hides are tanned—without the chemistry you will have . . . decaying skin.

Shira Leibowitz Schmidt: I hesitate to interrupt again, Roald, but I just remembered a chemical detective story about tannins. Primo Levi, the Italian chemist/writer (see the illustration below) wrote a unique autobiography, *The Periodic Table.*

In one chapter, Levi goes to a class reunion of the chemistry graduates of the University of Turin. He meets Cerrato, a chemist who had worked in the photographic industry in Italy and Germany, about whom Levi writes, "If I would allow him a professional metaphor, his years of study were his Technicolor, the remainder

Primo Levi. (Photograph by Bernard Gotfryd.)

was black and white." Cerrato arouses Levi's sympathy, and in a remarkable passage that evokes Levi's aims in writing this book (and resonates with ours), Levi says:[24]

> I told him that I was in search of events, mine and those of others, which I wanted to put on display in a book, to see if I could convey to the layman the strong and bitter flavor of our trade, which is only a particular instance, a more strenuous version of the business of living. I told him that it did not seem fair to me that the world should know everything about how the doctor, prostitute, sailor, assassin, countess, ancient Roman, conspirator, and Polynesian lives and nothing about how we transformers of matter live: but that in this book I would deliberately neglect the grand chemistry, the triumphant chemistry of colossal plants and dizzying output, because this is collective work and therefore anonymous. I was more interested in the stories of the solitary chemistry, unarmed and on foot, at the measure of man, which with few exceptions has been mine. . . .

Cerrato tells Levi such a story from his days working in a German factory manufacturing X-ray film. The title of the chapter is "Silver," and it is silver halide grains which are very sensitive to light as well as to other chemicals. The "photosensitivity" of silver halide grains makes them central to photography. Suddenly, the customers of Cerrato's company began to complain—their films were dotted with bean-size spots. The problem grew, there were threats of damage suits, the factory owners were frantic. Some detective work showed that all the defective photographic paper was manufactured on Wednesdays.

Still there was no clue of what was going wrong. Until Cerrato by chance spoke to a loquacious guard:

> One day he told me that he was a fisherman, but that for almost a year now he no longer caught any fish in the small river nearby: ever since they had opened a tannery five or six kilometers upstream. He then told me that on certain days the water actually turned brown. There and then I didn't pay attention to his remarks, but I thought about them a few days later when from the window of my room in the guest house I saw the small truck bringing back the overalls [of the X-ray film factory workers] from the laundry. I asked about it: the tannery had begun operating ten months before, and in fact the laundry washed the overalls in the water of the

stream where the fisherman could no longer catch fish. However, they filtered it and made it pass through an ion exchange purifier. The overalls were washed during the day, they were dried at night in a dryer, and sent back early in the morning before the plant opened.

I went to the tannery: I wanted to know when, where, how often, and on what days they emptied their vats. They sent me packing, but I returned two days later with the doctor from the Sanitation Office. Well, the largest of the tanning vats was emptied every week, on the night between Monday and Tuesday. They refused to tell me what it contained, but you know very well, organic tans are polyphenols and there is no ion exchange resin that can trap them, and what a polyphenol can do to silver bromide even you who are not in the field can imagine. . . . We tested it with increasingly diluted solutions, as homeopathic doctors do: with solutions of about one part to a million; we obtained bean-shaped spots, which, however, appeared only after two months of rest. The bean effect—*Bohneffekt*—had been reproduced in full: when all was said and done, it became obvious that a few thousand molecules of polyphenol absorbed by the fibers of the overalls during the wash and carried by an invisible piece of lint from the overall to the paper were enough to produce the spots.

Levi closes the chapter in his inimitable style, saying that he and his friend would keep in touch, and

each of us would gather for the other more stories like this one, in which stolid matter manifests a cunning intent upon evil and obstruction, as if it revolted against the order dear to man: like those reckless outcasts, thirsting more for the ruination of others than for their own triumph, who in novels arrive from the ends of the earth to thwart the exploits of positive heroes.

Roald Hoffmann: Look who's talking about digressions! But you can interrupt me any time, Shira—if it's for Levi or a chemical story.

Those tannins made a big difference in the Levi episode. They also play a role in differentiating leather (soaked in tannic acid) from parchment (bathed in lime). Another major difference is mechanical—the animal pelt that will become parchment is stretched tightly while it dries. The network of collagen fibers changes significantly as a result; under the tension some collagen links are broken, the remaining ones align into layers parallel to the surface. The dried sheet is smooth, strong, less flexible than leather, and lighter in

The parchment maker scrapes the stretched calfskin with a half-moon shaped, sharp-edged tool until it is thin and smooth. (Photograph by Leila Avrin.)

color. The illustration above shows the tawing of hides, on the way to parchment, in our times.

Rabbinic authorities were concerned that inscriptions be permanent. They forbid, for example, treating parchment with saffron. Today we know that saffron then contained alum, which closes up the pores of any skin and may leave residue which would later break down ink as it dried.[25]

Which brings us to inks. A typical Egyptian recipe for black carbon ink called for scraping the soot off cooking pots, mixing it with a weak solution of gum arabic (*Acacia arabica*), drying it in cakes, grinding it, and dissolving in water as needed. Ink such as this scraped or washed off pretty easily—as we have mentioned in the discussion of the ordeal of bitter waters, the trial of the suspected adulteress (in Chapter 4, "Bitter Waters Run Sweet"), where the words of her accusation were washed off into water for her to drink.[26]

While carbon inks washed away, metallic ones could be erased only by scraping. Permanency is primary. The fact that the opposite is a devastating curse ("may his name be erased") reflects the importance of writing, lasting writing, in Jewish culture.[27] Over the centuries, the method of ink-making in the surrounding cultures influenced the way it was made for Jewish ritual writing, and today scribes' ink is made from tree sap, gall nuts, and bronze and copper sulfate (the latter being metallic).[28] Gall nuts provide tannic acid for ink; combined with iron salts and oxygen the mixture turns black or brownish gray. The ink oxidizes or "burns into" the parchment (the word ink comes from *incaustum*, burning in).

I need to say something about paper, my favorite writing material. I have a romance with fine paper (and fountain pens). When last year the production of a certain light blue stationery was discontinued (I had been introduced to it by one of my teachers, R. B. Woodward, and then used it for decades), I was disconsolate. Paper has responsiveness, a personality. I can only compose on paper, with pencil or ink.

Paper is made up of small discrete fibers of vegetable origin, often chemically treated in some way, pressed, and dried. Papermaking originated in China around 100 C.E., spread to Samarkand about 750, and then gradually to the Near East and via Spain and Italy to the rest of Europe (by the thirteenth century). First flax, linen, hemp, and eventually cotton fibers were used. The popular innovation of linen underwear for peasants in the fourteenth century gave papermaking a boost (and according to James Burke, eventually led to Gutenberg's printing press of the 1450s).[29] When the linen undergarments wore out they were sold to the rag-and-bone collector for use in making paper. Many Jews did this (for few were Rothschilds . . .); there is even a good Yiddish expression for the profession—*shmatteh shlepper*. A later anonymous verse goes

> *Rags make paper.*
> *Paper makes money,*
> *Money makes banks,*
> *Banks make loans,*
> *Loans make beggars,*
> *Beggars make rags.*

Since the mid-1800s wood pulp has been the main raw material in paper. The most important component of pulp is cellulose, a biopolymer we have encountered before (see Chapter 1, "Is Nature

Natural?") as the precursor of rayon. The production of pulp appropriate for making a variety of modern papers is very intensive in chemical use—the pulp often has to be bleached with hydrogen peroxide or sodium hydrosulfite. Even before the nineteenth century, bleaches were used, for rags were colored or dirty! Chlorine gas and hypochlorite are also used in bleaching; the concern about chlorine-containing molecules (the vast amount of chlorine in natural salt notwithstanding) in our environment is leading to greater use of oxygen, ozone, and hydrogen peroxide as alternatives for bleaching. Various noxious, malodorous sulfur compounds such as hydrogen sulfide (H_2S), methyl mercaptan (CH_3SH), oxides of sulfur (SO_x), and dimethyl sulfide (CH_3SCH_3) may be released in paper manufacture. It's a struggle to reduce these emissions and make paper mills good neighbors.

Notwithstanding parchment for ritual uses, is there any question that paper has won the day as a writing and printing material? It is cheap, and comes from a renewable resource. And the spread of computers in industrialized countries, contrary to speculation, has only increased the demand for paper, more and more of which is recycled.

Shira Leibowitz Schmidt: Having an understanding of some of the components that go into the making of parchment and ink, we might ask: At what point in the making of a Torah scroll does it become a scroll with such sanctity that reading from it can fulfill the religious obligation to do so? *Which molecule has the sanctity?*[30]

To answer this you might come to Netanya, a half-hour north of Tel Aviv. In the apartment complex where I live in Kiryat Sanz, a *hasidic* enclave overlooking the Mediterranean, there are several dozen scribes who work out of their own homes or in groups in workshops. You could view some stages of the making of a Torah scroll, though you would have to follow it for a year to see the completion of an entire scroll.

My neighbor, Rabbi Reuben Dembs, learned the craft in an intensive course for scribes, so that he could supplement the income of his large family. He travels to a suburb of Tel Aviv to buy sheets of rolled parchment (which are prepared a good distance from the residential area, remember the pollution that might be involved), made from uterine calves, and special scribes' ink.

Rabbi Dembs buys some turkey feathers from which to whittle quills. It took him several weeks just to learn the art of carving the proper double-pointed quill needed to make the broad strokes, thin lines, and crowns of the special Torah letters. He sits in a corner of his porch, two of his eight children playing quietly nearby. A piece of near-white parchment is unrolled on the slanted table top. Faint guidelines for writing straight lines are ruled on the parchment by the scribe with an awl or ice pick, or are preruled in the parchment factory with a computer-aided device. Rabbi Dembs dips his quill into the ink and delicately forms each letter individually. There are approximately 250 columns of text in a Torah scroll, about 50 skins of parchment have to be sewn together. This is not done, however, until all the individual parchment sheets have been written. Then he goes to a neighbor who sews them together with sinews from a kosher animal. Each end of the long scroll is attached to a roller and the entire scroll is then rolled up. There are women in the neighborhood who sew scroll covers from velvet and embroider them with gold or silver thread; and there is a Yemenite silversmith nearby who is noted for designing the crowns, pointers, and breastplates.[31] The written parchment scroll alone costs about $25,000.

Roald Hoffmann: Now at what point in this process does the Torah acquire the sanctity that will enable a man to read the scroll in public, with the appropriate blessings, anywhere in the world—be it in the main synagogue in Jerusalem or in an army tent near the 38th parallel in Korea?

All the parts can be put together perfectly, but the scroll will not be fit, or "kosher," if the scribe himself did not inscribe each letter, each word, with the intent that it be used for sacred purposes. Before writing, the custom is for the scribe to immerse himself in a ritual bath (similar to the one shown on page 224, but for men). And, importantly, every day before commencing work, he makes a declaration committing himself to the holy purpose of his work. Still another dedication is said by him every time he writes any of the names of God. Thus, it is the focusing of his intention[32] that imbues the scroll with its sanctity while it is being written. As you told me, Shira, Rabbi Dembs says that it is not enough to think so, one must say it.[33]

Shira Leibowitz Schmidt: My family and I went last Tuesday to a festive ceremony celebrating the installation of a Torah scroll in Kiryat Sanz. Mrs. Esther Shapiro, my neighbor, commissioned the writing of a Torah as a living memorial to her recently departed father. On the anniversary of her father's death the ceremonious completion of the scroll writing took place. My husband, Baruch Schmidt, was included among those invited to inscribe one of the

Rabbi Zvi Elimelech Halberstam, Rebbe of the Sanz *Hasidim*, Netanya, at a Torah scroll completion and installation celebration. The Ashkenazi scroll he dances with is smaller than usual because the *rebbe*'s father (who survived concentration camps, resettled in the United States and then in Israel) preferred the Torah to be as portable as possible. (Photograph by Baruch Schmidt.)

last letters. He, a Texan transplanted to this *hasidic* world, had been warmly accepted into the religious life of the neighborhood.

As the men crowded into the Shapiro apartment where the nearly complete scroll lay, each one chose a letter to fill in. The last few verses had been outlined by the scribe so that each honoree had only to fill in the ink. Baruch dipped the turkey quill into the ink, but then, despite his dexterity as a surgeon, he declined. Saying "we learned in medical school, *'Primum nil nocere'*—first, do no harm," he begged off. If a mistake be made, the process of scraping the ink off is delicate (remember the ink is chemically composed to resist being worn off or erased). He asked the scribe to be his proxy for his letter (a Hebrew "S").

At nightfall my nine-year old, Michael, attired in his Saturday best, joined the hundreds of boys who formed a torchlight procession to accompany the Torah from the apartment around the community to the synagogue. Chagrined that only the eleven-year-olds were allowed to carry blazing torches in the parade, he joined the chorus singing psalms and dancing. Four strapping teens held a bridal canopy aloft over the Torah as it was danced through the streets, cordoned off by police for the festivities. [See photograph on the facing page.]

My Efrat, a few years older than her brother, was involved in the pastry-making and consuming (more of the latter). Yes, the gender-role distinctions are fairly clear-cut, and from a distance they may look sexist. But here the people take pride in what appears to be old-fashioned—ancient is beautiful, old is honored. As it says in the Mishnah Avot, "old wine in new flasks."

Roald Hoffmann: That reminds me, Shira, of what I thought was a puzzling juxtaposition of old and new in the Gospel of Luke, which I found in thinking about the Christian resonances of the title of our book. Luke reports Jesus, saying:

> And no one pours new wine into old wineskins. If he does, the new wine will burst the skins, the wine will run out and the wineskins will be ruined. No, new wine must be poured into new wineskins.

And then there comes the line:

> And no one after drinking old wine wants the new, for he says "The old is better."[34]

The last line appears only in the Gospel of Luke, not when the same parable is related in the other Gospels, and apparently not in all early manuscripts of Luke.[35]

I was really puzzled by this line, for if the metaphorical meaning of the parable was that Christianity was meant to replace Judaism, then this line seemed to negate what came before it. I turned to a friend, Jonathan Bishop, who is a New Testament scholar, and he said "Well, of course Luke is sarcastic. Just like him, more anti-Jewish than Mark or Matthew. He keeps saying things just like that, meaning 'Ha, those old rabbis—they've got it good, you can't expect them to change their ways!'"[36]

You've just said something similar, Shira, and I like it.

Shira Leibowitz Schmidt: Again you're interrupting, Roald, that male habit. *And* digressing.

Once inside the synagogue, the Torah was placed in the ark to the accompaniment of more singing and dancing. Mrs. Shapiro, who was watching from front row center in the women's balcony, tears on her psalter, was surrounded by several dozen proud grandchildren. Efrat, jaded by having been to several Torah installations, questioned, "Why the fuss?"

I pointed out to her that not every faith community centers its focus on a book; people do have other options such as art, music, and architecture. There are religious systems that speak through media other than books, and I reminded her of the summer we spent on the Navajo reservation in New Mexico when Baruch worked at the Ship Rock Hospital, opposite the Ship Rock, rising 1,800 feet off the desert floor. (See Plate 20.) The Navajo, a religiously sensitive people, had no written language and thus no books until recently. Yet they preserved and transmitted their tradition through other media—song, storytelling, and crafts. Their sandpaintings are both artistic and religious ceremonials, containing hundreds of symbols memorized by gifted artists through generations.[37]

Music can be a focus of religious experience: Bach's "Last Chorale" is a vehicle for profound religious experience, voicing in its intricate musical repetitions, their number and key, a statement of Protestant theology.

Often architecture can point us to the central values of a group. When we entered an ancient Anasazi (pre-Navajo) ruin, the circular ceremonial *kiva* enveloped us and focused our eyes on the religious wall paintings. Inside the chapel of the U.S. Air Force Academy in Colorado Springs, you feel a heavenward uplift through the arrangement of arches and pillars, as your eyes move vertically up the orchestrated sets of organ pipes.

Where does your eye focus when you enter a synagogue? The ark containing the Torah scroll is the central architectural organizing feature. But in addition it represents the entire endeavor of Torah study. The sanctity is not in the molecules of protein in the parchment, nor in the gall nuts of the ink, nor in the sinews of the stitching. The sanctity begins in the intention and dedication of a learned scribe, and is sealed in the labor and love of learning Torah.

Here's what one of the first modern Hebrew poets wrote:

> *If you wish to know the well*
> *from where your brothers going to their slaughter drew*

in days of torment the strength
to meet grim death with joy. . . .
If you wish to know the well
from where your stricken brothers drew
through hellish pains
trust in Heaven's comfort. . . .
Oh brother! if you do not know all these—
then betake yourself to the old study house,
in the sultriness of day, the dark of night
if there be left any remnant there
perhaps your eyes will discern. . . .
Jews who bear the weight of exile
forgetting toil in Talmud pages worn,
worries washed away in midrash,
pain relieved by Psalms. . . .

Hayyim Nahman Bialik[38]

Roald Hoffmann: I understand this, Shira. The Torah scroll, as a physical object, is also the crystallization of a much broader concept of Torah which is crucial to understanding Jewish tradition. Central as it is, the public reading of this handwritten scroll is only a small part of Torah. Here is what Jacob Neusner says:[39]

> In Judaism . . . if you want to meet God, you open a book and study. That labor of learning, studying a document of the Torah, is the first . . . most important place [to meet Him].

Shira Leibowitz Schmidt: This Torah comes in two media. One of them is the inscribed Pentateuch, so much the subject of this chapter. The other is the corpus of oral, and subsequently written, law, commentary, and responsa. These are the classics of the Jewish people—rooted in tradition, yet changing over the millennia. We have cited the tiniest fraction of them in this book.

> God is present in the Torah, in the act of study, and in what is studied . . . when you know that fact you know everything important about Judaism. . . .[40]

We have seen in the foregoing seven chapters there are many ways that the Jew meets God:

1. in the restrictions on people who pollute the environment;

2. in a *sukkah,* eating the holiday meals;

3. in the *tefillin,* wound on the left (or right!) arm and forehead;

4. in the hidden or not-so-hidden miracle of sweetened waters;

5. in the biblical blue dye from snails;

6. in the covenant that the rainbow symbolizes; and

7. in the laws of "purity" restricting food intake and marital relations.

But "the study of Torah is equivalent to them all."[41]

Roald Hoffmann: The Sanz community, in which you now live, Shira, was reconstituted from the remnants of European Jewry after World War II. Its continued existence, and its devout concentration on the Torah, bring us full circle back to the beginning of this chapter— to the Jews and their books.

In Chapter 2 we mentioned the wartime responsa of Rabbi Ephraim Oshry of Kovno, Lithuania. Here is one on the sanctity of a Torah scroll, entitled "The Sunken Torah Scroll":

> Immediately after the liberation, when G-d opened the gates and redeemed us from our bondage, a respected Lithuanian Jew poured out his heart to me. He related how he had seen the Germans bait an old rabbi and, under pain of death, force him to burn a Torah scroll with his own hands. Since the witness owned a Torah scroll himself, he began to worry lest they compel him to burn his Torah scroll too. He woke up in the middle of the night, took his Torah scroll to the river and let it sink to the bottom to keep the accursed Germans from desecrating the sanctity of the Torah.
>
> Nevertheless, he was troubled by doubts: Why hadn't he found another way to hide the Torah scroll? And who could say for certain that the Germans would have compelled him to burn the Torah scroll? And even if they would have compelled him to do exactly what they had forced the old rabbi to do, by what right had he assumed that he was allowed to throw the Torah scroll into the depths of the river before the problem actually came up? And which was more disgraceful—throwing the Torah scroll into the river or burning it?

A scroll of Esther being written in the hand of Rabbi Reuben Dembs. (Photograph by Kenneth L. Fischer.)

After liberation, the first thing he did was hurry down to the river in the hope of dredging up the Torah scroll from the bottom so that he could bury it in the traditional manner. But his efforts were in vain; he could not locate it. Then he came to ask whether he needed to atone for his action.

Response: When this man had cast the Torah scroll into the river, he had done so out of respect for its holiness and a clear desire to save it from further abuse. He was therefore not obligated to atone for his deed. But I suggested that when G-d in His bounty, granted him prosperity, it would be appropriate for him to set aside money and purchase a Torah scroll through which he would sanctify G-d. . . .[42]

Is it any wonder that Muhammad, the founder of Islam, called the Jews "The people of the Book"?[43] I began life as Roald Safran, before I became Hoffmann. The root of my family name (found in both Ashkenazic and Sephardic communities) is *s-f-r*, from which *sefer*, the word for book, comes.

Epilogue

We have come to the end of our modest effort to look at issues of science and Jewish religious tradition. Here's what you **will not** have found in what we have written:

- That religion is superstition, science is gospel, and the twain are fated to war with each other.
- That modern science supports/refutes the word of the Bible.
- A seductive identification of God with the edge of the universe, or for that matter with all the infinities of our ignorance.
- A scholarly causal explanation for *why* science and religion have commonalities.

To us the resonances in the stories we relate suffice—they are profoundly evocative of the spirit of both religion and science. We are satisfied (and amazed) by the cabbage looper moth's blend of pheromones set side by side with Rabbi Geffen's analysis of the blend that constitutes Coca-Cola. We are intrigued by young Pasteur's imperative to separate his mirror-image molecules before the eyes of Professor Biot, in that process probing the role of authority in science. Authority is also questioned in quite another context, religion, during the calendrical disagreement of Rabbis Gamaliel and Yehoshua.

Here is what you **will** have read:

- Juxtaposed tales of science and Jewish religious tradition, in which scientists, doing experiment and theory, face concerns similar to those of human beings in the Bible, the sages of the Talmud, and thoughtful people throughout the ages.
- A parallel not only in subject matter, but in the logic applied. There is a set of hypotheses, precedent, a way of

thinking. Yes, matter matters. And so does consistency within a framework of understanding. There is creation of ideas and molecules, but that creation is based on the work of others (sages, researchers, scholars all). There is incredible continuity, and clever, clever thought.[1]

Science indeed deals with the ordinary—the wonder in that suffices—but it will not remain in the ghetto of the material. To quote Richard Powers, science is

> [a] way of looking, reverencing. And the purpose of all science, like living, which amounts to the same thing, was not the accumulation of gnostic power, fixing of formulas for the names of God, stockpiling brutal efficiency, accomplishing the sadistic myth of progress. The purpose of science was to revive and cultivate a perpetual state of wonder. For nothing deserved wonder so much as our capacity to feel it.[2]

And religion—well, it simply refuses to bury itself in the study of wondrously rich tomes of the Talmud, the warm tones of the Sabbath.

Science and religion are both ways of trying to understand the world, to find meaning in that world's beauty and terror. Our stories are testimony to the struggle to understand. They bear evidence to the human condition—that people seek meaning, and people act. In these stories of matter and spirit, both Adams (and women too) speak, try to understand, exercise compassion, find the new. The voices of science and religion intertwine.

How shall we proceed, how shall we create and understand, how shall we find the new? For guidance, we return to Zloczow, and the words of Rabbi Yehiel Mikhal (who was roughly a contemporary of Antoine Laurent Lavoisier):

> The *maggid* of Zloczow was asked by one of his disciples: "In the book of Elijah we read: 'Everyone in Israel is duty bound to say: When will my work approach the works of my fathers, Abraham, Isaac and Jacob.' How are we to understand this? How could we ever venture to think that we could do what our fathers could?"
>
> The rabbi expounded: "Just as our fathers invented new ways of service, each a new service according to his own character: one the service of love, the other that of stern justice, the third that of beauty, so each one of us in his own way shall devise something new in the light of the teachings and of service, and do what has not yet been done."[3]

How We Came to Old Wine, New Flasks with a Little Help from Our Friends

Roald Hoffmann's version: In 1989 I traveled to Beer Sheva, Israel, to receive an Honorary Doctor of Science degree from the Ben-Gurion University of the Negev. The seasoned recipient that I am, I steeled myself to still another introduction. Rector Abraham Tamir surprised me, not only saying the things that pleased my mother and other family there, but also finding intelligent parallels in my life to that of Roald Amundsen, the Norwegian explorer after whom I was named. Someone had done some research, for a change.

Afterward I complimented Professor Tamir on his introduction. He said: "I cannot tell a lie. My assistant, Shira Leibowitz, wrote that introduction."

That evening I spoke for the first time to Shira. She also came to a lecture of mine on the theme of "Natural/Unnatural" the next day, a lecture for which she arranged an unusual preface, a performance of a segment from the *Tales of Hoffmann*. Obviously, Rector Tamir was very lucky in his choice of staff.

Shira listened carefully to what I said in the lecture—of the underlying molecular nature of both the natural and the synthetic, of the infinite ways in which chemists confound this separation, and of all the psychological and spiritual reasons why we nevertheless favor the natural. She remarked, "You know, in Jewish religious thinking there is something related going on. Take the materials for construction of a *sukkah*—the sides of the *sukkah* can be made of anything, but the roof of a *sukkah* only from very specific materials—plants once living, no longer alive, and not too much transformed by human beings. The rabbis had a long discussion on this fifteen hundred years ago."

I made the natural response of a writer to Shira's suggestion of a religious dimension to the natural/unnatural dichotomy—I said, "That's interesting, Shira. You should write something about this."

I had noticed from her attire that Shira was observant. I also knew from her accent that she was originally from the United States. In time I learned that her Stanford engineering degree didn't leave her with much chemical knowledge, that her husband Elhanan was a distinguished mathematical physicist, from one of Israel's prime intellectual families; that she (as many in the Leibowitz clan) was that *rara avis*—highly observant yet politically liberal (this was Israel and politics matters a great deal), and that she was the mother of six, a full-time job. She was also a very careful reader and listener, taping lectures she liked and listening to them incessantly, critically. Shira did not like to waste a moment, and she was inordinately fond of lists of things to do.

When I called Shira a talmudic scholar, she vehemently objected to the authority I had thus conferred—she had studied a little, she said, that's all. Women do study the whole range of sources today, and her husband's aunt, Nehama Leibowitz, of blessed memory, was a legendary teacher and commentator, a person acclaimed (by those religious who can accept the notion that a woman can be such) as quite probably the most perceptive Bible scholar of her day.

I then went back to the United States and forgot about Shira's comments. Shira didn't. She obviously needed someone to prod her to write (the Leibowitz family—her own, her father-in-law's—could be a pretty intimidating setting for intellectual activity). Shira saw the holiday of *Sukkot* coming up, and so she wrote up her ideas (now in the second chapter of this book) about just that discussion of natural/unnatural mentioned above, in the context of the materials science of a *sukkah*. And she conceived the idea of trying to publish that essay, as a seasonal piece, in the *Jerusalem Post*.

Shira sent the piece to me, saying it really had to be submitted yesterday (here I learned she just loves deadlines) and put both our names on it. I didn't like the hurry, and so I said, "Shira, why don't you send it in under just your name?" Reluctantly, she agreed. I then read the essay and saw that in part Shira had just summarized, nearly verbatim, material from my lecture. And I liked the way she integrated the religious material, with verve and intelligence. So I called back and said, "I have no shame. I don't want you accused of plagiarism. If it's still O.K., you can put my name back on the article."

It was not too late, and so in fall 1989 *The Jerusalem Post* published our first article. The venue was appropriate, in that one does not expect great scholarly work published in a newspaper. From the beginning (and this book is the ultimate proof) Shira and I knew we were not scholars. Often we've relied on secondary sources. We were simply interested in the fascinating interconnections of religious and scientific ways of looking at the world. And we liked writing.

Over the next three years we wrote three further articles, two of which ("Signs and Portents" and "Pure/Impure") are in this book in modified form. We were lucky to find a home for them in prominent American literary and critical journals—the *Michigan Quarterly Review, Diacritics*, and the *New England Review*. This work was accomplished by mail (and not yet by e-mail); we met actually only briefly, for one day, in the course of our writing.

Encouraged—naively—by the ease of collaboration with Shira so far, and by some simplistic notion that the scientist's mastery of the ethic and process of collaboration (almost all my professional papers have several coauthors) could be transferred to creative work in the arts, I conceived the idea, and Shira did too, that maybe we might collaborate on a book.

As we worked on the articles, it emerged that Shira, who was very religious, had great respect for science (and liked chemistry, even though she still refused to learn it). And I, who was (and am) unreligious, had great respect for religion in general, and for our Jewish religious heritage specifically. I have a feeling for religious thinking, and for ritual. When I read the words of the morning service, I am with the people, the words mean much to me.

At times, I think Shira might have mistaken this for a sign that I might be "convertible," that is, might return to the observant Jewish fold. But then I disabused her of that notion by an occasional antireligious outburst, or a citation of a religious source that is not Jewish. Just a few of which have survived in this book. Shira might have been less likely to think I was potentially observant if she knew that the same participatory feeling comes over me when I hear a mass sung in a Catholic church.

Respect for science and religion, some tension from the conjunction of the twain, a communality in being Jewish, yet still another tension from one of us being observant, the other not, the natural duality of male and female viewpoints, a common interest in the arts, a common disdain for all those books equating God with everything we don't know . . . these are the sources of our collaboration. They also led, quite naturally, to the dialogical structure of many of the chapters.

Sadly, Shira was widowed in 1990. Aside from the personal tragedy to her and her children, this event set us back in our work. Shira eventually remarried, and in Dr. Baruch Schmidt, a physician, found the support that has allowed her to go on with our work. In the selfish way we have of thinking of the world, my concern when I heard of her courtship and marriage was, "Can we go on with our writing?" The point is that in the orthodox world in which Shira lives (and I do know of its benefits, richnesses, and diversities) it is not a given that a wife coauthor books or articles of this nature. I could not assume we would continue.

We did! I will tactfully refrain from describing the ways in which the actual writing of this book has changed all of my ideas about collaboration. I'll change my mind again. . . . I will say, though, that I am grateful to Shira Leibowitz Schmidt for leading me into the world of Bible, *midrash*, Talmud, commentaries, responsa. It's a far richer world, that of the religious traditions of my people, than I had ever imagined. I love it.

Probably some people who come to this book will think that I am more responsible for the ideas and Shira for the research (such as it is) and for the writing. That actually is *opposite* to the way our collaboration has evolved, to the extent one may apportion such things. It is Shira who has most of the ideas. It is she, creatively stubborn, who persists with those ideas, sneaking them back in when I have tried to dismiss them. And I admit it, I'm mainly responsible for the writing, because writing comes easily to me.

But we won't tell you which parts that appear in one author's name were actually written by the other!

Shira Leibowitz Schmidt's version: It was 3:15 on a hot, desert afternoon in Israel. The notice for the lecture said:

Title: Natural/Unnatural
Guest Speaker: Prof. Roald Hoffmann
Nobel Prize Chemistry 1981
Auditorium 03, May 21, 1989, 3 P.M.

The lecturer was being introduced by the rector, Professor Tamir of Ben-Gurion University. A melange of latecomers outside the door peered into the auditorium. They were perplexed when they saw on stage a soprano singing the "Barcarolle" from Offenbach's *Tales of Hoffmann.* They checked the auditorium number on the door against the notice. A scruffy student started to enter, but a sandaled professor protested: "No, this isn't it. We're looking for a chemistry lecture, not an opera. Must be a mistake in the auditorium number." Another student suggested they enter anyway. The professor prevailed and they went hunting across campus (in vain) for the lecture.

Unfortunately, those latecomers missed a wonderfully illustrated presentation on how the natural and the synthetic differ, if indeed they do. As a lecturer in English for science students, I had been asked by the rector to write something introducing Roald Hoffmann. I wrote an introduction using as a framework Offenbach's nineteenth-century tales of the three women Hoffmann loved. But in my adaptation the first of the loves was chemistry, the second was poetry, and the last was a promising and brash nineteen-year-old . . . Ben-Gurion University, Israel's youngest. The well-known aria capped the presentation.

Roald went on to discuss why we prefer the natural to the human-made. Afterward I suggested to him that the distinction between natural and unnatural, though fuzzy as he had shown, applied to at least two areas of Jewish life: the leafy covering of the *sukkah* booths during the Feast of Tabernacles, and the wearing of natural-looking wigs by religious women to cover their own hair. He suggested that I write up my thoughts, which I subsequently did. This led to our first publication, "Can You Build a *Sukkah* from an Elephant?" (and several years later to research on wigs for "Is Nature Natural?")

About that time, Roald asked me what I knew about the biblical blue, *tekhelet*, which was the subject of a book his family friend, Ehud Spanier, had edited. I knew the book, but was more interested in how Jewish law and lore treated this special dye than in the chemistry. I could see then that there was a sequence of events that we could trace from the biblical command to wear a fringe of blue, through the chemistry of dye-making, up to the colors of the Israeli flag today. Despite Roald's persistent dismissal of my suggestion we write a play about it, he eventually came around and even became quite an enthusiastic coplaywright of "The Flag That Came out of the Blue."

"Everything is related to everything else." Some chapters spawned others; the work on *tekhelet* led circuitously to the semiotics of "Signs and Portents." In looking at what contemporary scholars had written on the symbolism of the biblical blue, I came across the writings of Nehama Leibowitz, my late aunt-in-law. She had posed the question of whether one could use Nahmanides's dual interpretation of the rainbow to understand the meaning of the blue dye. I could not see any connection. "I'm stumped by your suggestion that the two are related," I complained to her one Sabbath afternoon. Leaving the five older children with my late husband Elhanan, his aunt and I pushed my infant son in a stroller through the quiet alleys of Jerusalem. As we walked she elaborated on Nahmanides's paradigm. To make her point clear she asked me to explain to her the meaning of each traffic sign that we passed (she didn't drive). As I interpreted them one by one, she pointed out that each could be classed as a concrete or abstract symbol à la Nahmanides, just as the biblical blue could be similarly categorized.

When several months later a police officer apprehended me because I misapprehended a misaligned no-parking sign, I tried to defend myself with Nahmanides's commentary on the rainbow. "Go tell it to the judge," he said, and I did. This provided the framework for the courtroom drama, "Signs and Portents," in which Roald and I could bring to bear our various interests: his in art and chemistry, mine in linguistics and Jewish symbolism. Though I lost the case we gained a chapter.

The search for an understanding of the biblical blue took us in other directions, eventually to the rocky Mediterranean shore of Israel where we snorkeled for the elusive *murex* snail (under the direction of Amutat P'til Tekhelet). Though I came up empty-handed, my children and husband, along with Roald, found snails which we processed into dye then and there on the beach in the same dye pit that had been used centuries before for this purpose. We dipped white wool into the snail secretion and watched the reenactment of that ancient dye dance from yellow, brown, mauve, to the finale of blue.

Inspiration for other chapters came from many quarters:

A quip a lecturer made in a Bible class on the bitter waters of Marah, "God may have been giving Moses some chemistry lessons," sent me off to discover what I could about whether the sweetening of the waters was a natural or miraculous transformation. And that set Roald off to check out the various chemistry processes that could have changed the bitter waters to sweet.

The involvement of my husband, Baruch Schmidt, in perpetuating the hand-papermaking workshop established by his late wife, along with Roald's abiding obsession with fine fountain pens led me to the story of the paper, pens, parchment, and camel caravans, which ends the book.

The children did their share of inspiring. One morning I told my (then five-year-old) Efrat to put on her blue sneakers. "Mother, those aren't blue. They're *tekhelet*." "Oh, what's the difference?" I queried. "*Tekhelet* is prettier," I was instructed.

While working one summer in Ithaca on "You Must Not Deviate to the Right or the Left," I was asked by Rabbi Eli Silberstein to give a talk on the Sabbath about our work. "And be sure to include some art!" With only a few hours advance notice, I searched in a panic to find a related painting, and found one just in time. With that lead, Roald subsequently discovered over a dozen portrayals over the centuries of the same subject matter: Jacob crossing his left and right hands while blessing his grandsons. The more difficult hurdle was to narrow it down to a single choice. One of us favored the Rembrandt; the other, the van der Werff. Finally the Rembrandt lobby gallantly relented—this time around—and the result is the portrait by van der Werff in "You Must Not Deviate to the Right or the Left."

As icing on the cake, we discovered that this painting was one of the earliest in art history to use the chemically synthesized Prussian blue pigment—a color that played a role in the history of the reintroduction of *tekhelet* that we describe in "The Flag That Came out of the Blue." This takes us full circle back to the biblical blue, near the start of the collaboration. . . .

And now I hope we can get started on a sequel to *Old Wine, New Flasks* so we can give the Rembrandt a chance this time.

The Help (More Than a Little) from Our Friends: Before we started the book we had to face questions of support. applications to the National Endowment for the Humanities and the National Science Foundation found us falling in the predictable cracks between disciplines, as well as facing reviewers who wanted a scholarly book (the one they couldn't write). They'll be disappointed. Eventually we gave up, and found the requisite material support in Cornell University funds generously made available for Roald's use.

Nat Sobel, a dedicated agent, tried hard to sell our project for a book, and for this we are very grateful. Elizabeth Ziemska, an agent with Nicholas Ellison then took on the project, saw the same thing we saw in it, and with great enthusiasm and effort found for us W. H. Freeman and Company. There Deborah Allen has been the most interactive and helpful editor either one of us has ever known. We have also had tremendous help from the other talented people at W. H. Freeman—Martha Solonche, Maria Epes, Sloane Lederer, Blake Logan, Thomas Brown, and Allyson J. Siegel.

So many people have helped us. Foremost among them have been two undergraduates at Cornell, Glenn Halpern and Oren Scherman, who faced with equanimity and resourcefulness the most obscure of fact checking and textual questions. They have contributed much to the final shape of the book. Roald's research group also provided valuable assistance.

There are several persons who have read drafts of the entire book and offered valuable comments to us. These include the readers of the book for the publisher—Michael S. Berger, Harvey D. Luber, Fred Naider, Brian D. Shmaefsky—as well as Patricia Linn, Ida Ayala Myers, and Baruch Schmidt. Important comments on individual chapters, which have led to real improvements, have been made by Rabbi Shalom Carmy, Eva B. Hoffmann, Rabbi Eli and Chana Silberstein, and Ingrid H. H. Zabel.

We thank the editors of *Diacritics, The New England Review*, and *The Jerusalem Post* for their editorial assistance in the initial publication of versions of three chapters in this book.

We are grateful to many friends and colleagues for helping us with facts, illustrations, quotations, leads to readings, and so much more. These include: Alex Altman, Yehuda Amichai, Rabbi Selig Auerbach, Rabbi and Mrs. Uri Auerbach, Leila Avrin, Alfred Bader, David L. Beveridge, Jonathan Bishop, Donald Boyd, Ross Brann, K. C. Cole, Derek A. Davenport, Herbert Deinert, Rabbi Reuben Dembs, Cyril Domb, Barak Epstein, Albert Eschenmoser, Thomas Eisner, Michael E. Fisher, Judith Fineberg, Hillel Furstenberg, Shmuel Gilbert, Norman Goldberg, Andy Goldfinger, Moshe Greenberg, Syd Greenberg, Ari Greenspan, Emily Grosholz, Bruce P. Halpern, Neal Hendel, Irwin Hollander, Ronald Jacobowitz, Peter A. Ward Jones, Jane Jorgensen, Inge Bösken Kanold, Yaakov Klein, Wolfgang E. Krumbein, the late Nehama Leibowitz, C. Frederick Leydig, Rabbi Nathan T. Lopes Cardozo, Barbara Mantaka, Fredric M. Menger, Brenda Miller, Thomas Morton, Pinchas Osher, David Owen, Earl Peters, Bernard Rabenstein, Ralph Raphael, Ruth Reingold, Wendell Roelofs, Max Saltzman, Sason Shaik, Elizabeth Sherman, William Shirley, Marilyn Shnider, Jane Singer, the late Cyril Stanley Smith, Ehud Spanier, Nellie Stavisky, Baruch Sterman, David Thorn, Dennis Underwood, A. Varvoglis, Stephen J. Weininger, Edward Wolpow, and Israel I. Ziderman.

Kenneth L. Fischer is thanked for his expert photography, and Cornell University Photography for their patient and expeditious services. Gary Michalec helped us tremendously in composing and producing some of the images.

Roald takes ecumenical pride in the fact that pieces of this book were written at Yeshiva University and at the Pontifical Catholic University of Rio de Janeiro. He is grateful to both institutions for their hospitality, and especially to Maria Matos in Brazil.

Shira is grateful to Phil and Lesly Lempert for their hospitality in the course of her visits to Ithaca, and the members of the Jewish Living Center at Cornell University for providing a home away from home.

Shira thanks, above all, her husband Baruch and her pearls and rubies, Shamai, Nathaniel, Tirza, Akiva, Efrat, and Michael for their intellectual criticism, patience, and help in difficult moments.

Notes

* Note on transliteration: Source material transliterated by others has been left in its original form, even where it differs from our book's convention.

Preface

1. This story is based on the authors' paraphrase and translation of Midrash, Genesis Rabbah 8:5. The words of Truth and Peace come from Jacob Neusner's translation, *Genesis Rabbah, The Judaic Commentary to the Book of Genesis. A New American Translation* (Atlanta: Scholars Press, 1985), p. 78. We thank Rabbi Nathan T. Lopes Cardozo for helpful discussions of this *midrash*.

2. For a perceptive account of the many ways that science and religion relate to each other see Ian G. Barbour, *Religion in the Age of Science* (San Francisco: Harper, 1990).

Chapter 1: Is Nature Natural?

1. Genesis 1:28. Here Ayyal uses The King James translation [*The Holy Bible: King James Version* (New York: Meridian, 1974)], which has a familiar ring.

2. Job 38:24–31. Unless noted otherwise, throughout this book we use the translation of the Jewish Publication Society of America, and refer to it as JPS: *The Torah; The Prophets; The Writings: A New Translation of the Holy Scriptures According to the Traditional Hebrew Text* (Philadelphia: The Jewish Publication Society, 1962–1982).

3. One of the authors called (well, wrote to) "his" rabbi, Shalom Carmy, who pointed out that the root *t-v-a* appears in the chapter that Ayyal quoted earlier, Job 38. In verse 6 it has the meaning "sunk."

4. See H. Malter, "Mediaeval Hebrew Terms for Nature," in *Judaica Festschrift: zu Hermann Cohens Siebzigstem Geburtstage*, I. Elbogen,

B. Kellermann, and E. Mittwoch, eds. (New York: Arno Press, 1980), pp. 253–56.

5. For a general definition see I. A. Ben Yosef, *The Concept of Nature in Judaism* (Cape Town: University of Cape Town, 1986). There is also much interesting material on medieval Jewish attitudes toward nature in David B. Ruderman, *Jewish Thought and Scientific Discovery in Early Modern Europe* (New Haven: Yale University Press, 1995), Ch. 1.

6. Roald Hoffmann, "Natural/Unnatural," *New England Review and Bread Loaf Quarterly* (1990) 12(4): 323–35.

7. Arthur O. Lovejoy and George Boas, *Primitivism and Related Ideas in Antiquity* (New York: Octagon Books, 1973), pp. 447–56.

8. The Greek word *physis* (from which come *physics* and *physician*) had very clearly the second meaning of "nature." The first meaning of nature was covered by the all-encompassing *kosmos*. See also the excellent discussion in the Introduction to John Boswell, *Christianity, Social Tolerance, and Homosexuality* (Chicago: University of Chicago Press, 1980), pp. 11–15; also the highly readable essay by George Seddon, "The Nature of Nature," *Westerly* (December 1991): 7–14.

9. G. E. R. Lloyd, "The Invention of Nature," in John Torrance, ed., *The Concept of Nature: The Herbert Spencer Lectures* (Oxford: Clarendon Press, 1993), pp. 417–34.

10. Wendell Berry, "The Journey's End," in *Words from the Land: Encounters with Natural History Writing*, S. Trimble, ed. (Salt Lake City: G. M. Smith, 1988), pp. 226–38; quote from p. 230.

11. Arturo Gómez-Pompa and Andrea Kaus, "Taming the Wilderness Myth," *BioScience* (1992) 42: 271–79.

12. The Spanish manuscript *Haggadah* is in the John Rylands University Library of Manchester. The book Ayyal found is David Goldstein, *Jewish Legends* (New York: Peter Bedrick, 1988).

13. Paulus Potter, *The Young Bull*, oil on canvas, 235.5 × 309 cm, Mauritshuis, The Hague. There is a marvelous essay by Frank Stella that discusses this painting: Frank Stella, *The Dutch Savannah, De Hollandse Savanne* (Amsterdam: Stichting Edy de Wilde-lezing, 1985).

14. Part II, Chorus 26. The text is from Isaiah 53:6.

15. Richard Strauss, *Eine Alpensinfonie* (op. 64) (Munich: Leuckart, 1941). The section is called "Auf der Alm."

16. Brenda Miller (POB 281, Hod Hasharon, Israel), in her music seminars in Israel, has lectured widely on the development of the Nature theme in music and art. We thank her for the examples of the Handel chorus and the Strauss symphony.

17. The translators of the Bible (and of the Hebrew poems we will soon cite) take many liberties with the words for *deer, ram, hind,* and *buck.*

18. The familiar King James Version is used for Psalm 23, whereas the JPS translation was thought to be better for Psalm 42.

19. Ralph Waldo Emerson, *Nature* (Boston: J. Munroe, 1836); facsimile edition (Boston: Beacon, 1985), Ch. 1, pp. 12–13.

20. Angelo Rappoport, *The Psalms in Life, Legend, and Literature* (London: 1935), p. 40, as quoted in Nehama Leibowitz, *Book of Psalms* (New York: Hadassah Education Dept, 1971), p. 21. See also Gerald J. Blidstein, "Nature in Psalms," *Judaism* (Winter 1964), 13(1): 29–36.

21. The King James Version is used here.

22. Angelo Rappoport, *The Psalms in Life, Legend and Literature,* cited in endnote 20.

23. Ronald Paulson, *The Beautiful, the Novel, and the Strange: Aesthetics and Heterodoxy* (Baltimore: The Johns Hopkins University Press, 1996) maintains "the artist becomes a kind of surrogate God."

24. The complex and remarkable life of Wallace Carothers is described in a recent biography by Matthew E. Hermes, *Enough for One Lifetime* (Washington, D.C.: American Chemical Society, 1996).

25. For more on textile chemistry, see Kathryn L. Hatch, *Textile Science* (Minneapolis/St. Paul: West, 1993).

26. This section is adapted from Roald Hoffmann, "A Natural-Born Fiber," *American Scientist* (1997) 85(1): 21–22, where a recipe for an index of "naturalness" is suggested.

27. See Chapter 5, "The Flag That Came out of the Blue."

28. *Tzitzit* is a fringe. A *tallit* is a prayer shawl, with fringes in each of four corners. A *tallit katan* (or *arba konfot*) is a small *tallit,* a four-cornered poncho-like garment worn under outer clothing by Jewish males. A blessing is pronounced upon donning it. Just to make life difficult, often the term *tzitzit* is used to refer to the whole *tallit katan.* See illustrations on pages 59, 104, 160, and Plates 11 and 12.

29. J. David Bleich, *Contemporary Halakhic Problems,* Vol. II (New York: Ktav, 1983), pp. 50–53; also pp. 12–15 and pp. 43–45.

30. The story is more complicated. Rabbi Moshe Feinstein, who wrote the opinion Bleich cites, knew very well that the synthetic could be produced as a fiber. His reasoning (not undisputed by other authorities) was that the potentiality of extrusion of the synthetic material in sheets (the natural way in which leather comes) was sufficient to exclude the synthetic.

31. Joseph Bruchac, in Marie Harris and Kathleen Aguero, eds., *An Ear to the Ground* (Athens: University of Georgia, 1989), p. 46.

32. Roald Hoffmann, "Natural/Unnatural." *op cit.* See also Jeffrey L. Meikle, *American Plastic: A Cultural History* (New Brunswick, N.J.: Rutgers University Press, 1996) for a perceptive discussion of these issues.

33. Roger S. Ulrich, "View Through a Window May Influence Recovery from Surgery," *Science* (1984) 224: 420–21. We were led to this reference by a beautifully written article by Holmes Rolston, III, "Does Aesthetic Appreciation of Landscapes Need to Be Science-Based?" *British Journal of Aesthetics* (1995) 35(4): 374–86.

34. In addition to the sources about to be cited, see T. Carmi, *The Penguin Book of Hebrew Verse* (London: Penguin, 1966).

35. Translated by David Goldstein, *Hebrew Poems from Spain* (New York: Schoken, 1966), p. 64.

36. Translated by Raymond P. Scheindlin, *Wine, Women & Death: Medieval Hebrew Poems on the Good Life* (Philadelphia: Jewish Publication Society, 1986), p. 124.

37. Translated by Raymond P. Scheindlin, tr., *Masoret* (1995) 5(1): 3. Omitted lines indicated by a row of dots. For an analysis of the poem, see Raymond P. Scheindlin, "Poet and Patron: Ibn Gabirol's Poem of the Palace and Its Gardens," *Prooftexts* (1996): 16:31–47.

38. Raymond Scheindlin's perceptive commentary on this and many other poems deserves reading. He says that in ibn Gabirol's poem "the boundary between inanimate objects and living beings is deliciously vague." (See note 37, "Poet and Patron.")

39. Joshua 10:12.

40. Ross Brann, *The Compunctious Poet: Cultural Ambiguity and Hebrew Poetry in Muslim Spain* (Baltimore: The Johns Hopkins University Press, 1991), p. 158.

41. Translated by Raymond P. Scheindlin, in *The Gazelle: Medieval Hebrew Poems on God, Israel, and the Soul* (Philadelphia: Jewish Publication Society, 1991), p. 65.

42. R. Scheindlin, *ibid.*, p. ix.

43. Talmud, Ketubot 72a.

44. Talmud, Yoma 47a.

45. Numbers 4:11–31.

46. The rebellion is that of Korah, described in Numbers 16 and 17. See Chapter 5, "The Flag That Came out of the Blue," Act I, Scene 2, based on Numbers Rabbah 18:20.

47. Avigail Sheffer and Hero Granger-Taylor in *Masada IV: The Yigael Yadin Excavations 1963–1965, Final Reports* (Jerusalem: Israel Exploration Society, 1994), pp. 216–20.

48. Lisa Aiken, *To Be a Jewish Woman* (Northvale, N.J.: Jason Aronson, 1992), p. 133.

49. A perceptive reader, Patricia Linn, comments: "The point is to be attractive, seductive only to the husband, not to men outside home. But it turns the other way around. Women shave or cut their hair short for the husband and use nice wigs for others!"

50. The first Sanz Rebbe (b. 1793) didn't just have his family avoid wigs; he said in a responsum: " . . . G-d forbid that the women of our land should be permitted to follow this custom; the prohibition [against wearing wigs] stands." Rabbi Yitzhak Bromberg, *The Sanzer Rav and His Dynasty*, Shlomo Fox-Ashrei, trans. (Jerusalem: Mesorah, 1986), pp. 163–65.

51. Talmud, Baba Batra 60b.

52. Rabbi Mayer Schiller, "The Obligation of Married Women to Cover Their Hair," *Journal of Halacha and Contemporary Society* (Fall 1985): 81–108; Rav Moshe Feinstein, *Iggerot*, Responsa, Section 2:12 and Orach Chaim 112:4.

53. See illustration on page 217.

54. Dinel® and Elura® came into use for wigs after World War II; earlier, yak hair and corn silk were often used.

55. A sign near the Yaffa Wigs shop reads, in Hebrew, "It's natural to buy Yaffa Wigs!"

56. Talmud, Berakhot 24a.

57. An interesting, and to many puzzling, rabbinical regulation is that women are forbidden to sing individually before men other than their husbands. Some authorities even extend this ban to groups of singing women. See Saul J. Berman, "Kol 'Isha," *Rabbi Joseph H. Lookstein Memorial Volume*, Leo Landman, ed. (New York: Ktav, 1980), pp. 45–66. The article shows very well how post-Talmudic commentators shape a custom.

A memorial service for Yitzhak Rabin held in December 1995 New York City ran into trouble with certain orthodox groups when the initial plan called for Barbra Streisand to sing. She graciously backed out to avoid controversy.

58. Diane Ackerman, *A Natural History of the Senses* (New York: Vintage, 1991), p. 276.

59. Dr. Judith Fineberg, an observant radiologist, wrote in a personal communication:

> Actually what is considered erotic is somewhat dependent on culture. In the U.S. underarm hair is the pits and considered gross

by some, and women go to great lengths to depilate their legs. In contrast Europeans go *au naturel* and seem to attract men just fine. Admittedly hair on the head is a more universal come on. I cover my head since in our culture a woman's hair is so commonly seen, we are desensitized to its potential erotic quality (although I can't speak for a man). In addition, it is a Morse code: this woman is taken, STOP; this woman is available, GO GO GO.

60. Rabbi Pesah Falk, *Halachos and Attitudes Concerning the Dress of Women and Girls* (England: Gateshead, 1993), p. 6 says:

> It is however totally incorrect and against *razon hatorah* [intent of the Torah] for a woman to wear a *sheitel* [wig] that has been made to such perfection that it appears to the onlooker that this is the woman's own hair. . . . The *Gedolim* [great religious authorities] of *Eretz Yisroel* [Israel] have publicized that *sheitels* longer than shoulder length, loose wild-haired *sheitels* and highly inflated *sheitels* are below the standard of refinement that *tznius* requires. Unfortunately the *Yetzer Horah* [evil spirit] has caused impurity rather than *Kedusha* [holiness] to come from *sheitels,* and there are women who look "more girl-like" in their *sheitels* than true girls. This could not be further from the will of *Hashem* [G-d]. Anyone with a feeling and understanding for *tznius* is deeply offended by this trend. (Even names given to different styles of *sheitels,* e.g., Glamorous, are taken from the permissive and indulgent world we live in.) Rabbi Falk may have seen the Yaffa catalogue. It's good that the title of his book has "Attitudes" in it. . . .

61. From *Hair,* words by James Rado and Gerome Ragni, music by Galt MacDermot (Los Angeles: United Artists, 1979).

62. Ben Jonson, *Epicoene,* Act I, Scene 1 (New Haven: Yale University Press, 1971).

63. Albert Eschenmoser and Eli Loewenthal, "Chemistry of Potentially Prebiological Natural Products," *Chemical Society Reviews* (1993): 1–16; Albert Eschenmoser and Max Dobler, "Warum Pentose und nicht Hexose-Nucleinsäuren?" *Helvetica Chimica Acta* (1992) 75: 218–59.

64. This letter is adapted from Roald Hoffmann, "Unnatural Acts," *Discover* (1993): 21–24. We are grateful to *Discover* magazine for allowing us to reprint the text and illustrations.

65. This letter is a direct quotation from a comment on this subject by Barak Epstein. We are grateful to him for allowing us to use his words.

66. G. Blidstein, "Tikkun Olam," *Tradition* (1995) 29(2): 5–43.

67. From the beautiful essay by Wendell Berry, *The Gift of Good Land* (San Franscisco: North Point, 1981), pp. 267–81.

68. Lynn White, Jr., "The Historical Roots of Our Ecological Crisis," *Science* (1967) 155: 1203–07.

69. For some Christian responses to the White essay, see Wendell Berry, *op. cit.*, and S. Marianne Postiglione, R.S.M., ed., *Christianity and the Environmental Ethos* (St. Louis: ITEST Faith/Science Press, 1996).

70. "Biblical Take(s) on Environmentalism," *Johns Hopkins Magazine* (April 1996): 60.

71. Meir Tamari, *"With All Your Possessions": Jewish Ethics and Economic Life* (New York: Free Press, 1987), pp. 296–97.

72. Mishnah, Baba Batra 2:9.

73. See other references in Aubrey. Rose, ed., *Judaism and Ecology* (London: Cassell, 1992); *Judaism and Ecology* (New York: Hadassah, 1993); Leo Levi, *Torah & Science* (Jerusalem: Feldheim, 1983), Ch. 5; and Esti Dvorjetski, "Education for the Prevention of Air Pollution in Rabbinic Literature," in *Preservation of Our World in the Wake of Change*, Y. Steinberger, ed., Vol. VI A/B (Jerusalem: Israeli Society for Ecology and Environmental Quality Sciences, 1996), pp. 202–03; "Air Pollution and Public Health in the Days of the Second Temple, the Mishnah, and the Talmud," *Michmanim* (1998) 13, Hecht Museum, Haifa University.

74. *T'shuvot Maharshach,* part 2, subsection 98, as cited by Meir Tamari, *op. cit.*

75. As cited by Aubrey Rose, *op. cit.*

76. Joseph B. Soloveitchik, *The Lonely Man of Faith* (New York: Doubleday, 1965).

77. Other religious traditions have recently moved to consider environmental concerns explicitly. On January 1, 1990, Pope John Paul II begins his message "Peace with God the Creator, Peace with All of Creation" with a reflection on the biblical account of creation. After citing Genesis 1 and 2, the Pope goes on to say:

> Adam and Eve's call to share in the unfolding of God's plan of creation brought into play those abilities and gifts which distinguish the human being from all other creatures. At the same time, their call established a fixed relationship between mankind and the rest of creation. Made in the image and likeness of God, Adam and Eve were to have exercised their dominion over the earth (Gen. 1:28) with wisdom and love. Instead, they destroyed the

existing harmony by deliberately going against the Creator's plan, that is, by choosing to sin. This resulted not only in man's alienation of himself, in death and fratricide, but also in the earth's "rebellion" against him (cf. Gen. 3:17–19, 4:12).

After a discussion of the Christian beliefs on the role of the death and resurrection of Jesus, John Paul II goes on to say:

> These biblical considerations help us to understand better the relationship between human activity and the whole of creation. When man turns his back on the Creator's plan, he provokes a disorder which has natural repercussions on the rest of the created order. If man is not at peace with God, then earth itself cannot be at peace: "Therefore the land mourns and all who dwell in it languish, and also the beasts of the field and the birds of the air and even the fish of the sea are taken away" (Hos. 4:3).
>
> The profound sense that the earth is "suffering" is also shared by those who do not profess our faith in God. Indeed, the increasing devastations of the world of nature is apparent to all. It results from the behavior of people who show a callous disregard for the hidden, yet perceivable requirements of the order and harmony which govern nature itself.

The Pope's message continues with a wide-ranging discussion of the ecological crisis and of poverty as moral issues, and, interestingly, of the aesthetic value of creation. He ends with an exhortation to harken to St. Francis of Assisi's inspired "Canticle of the Sun."

78. Much of this letter is quoted, with permission, from a correspondence with Barak Epstein.

Chapter 2: A *Sukkah* from an Elephant

1. See *Moritz Oppenheim: The First Jewish Painter* (Jerusalem: The Israel Museum, 1983). The original German collection was *Bilder aus dem altjüdischen Familienleben nach Originalgemälden von Professor Moritz Oppenheim*, with an introduction and explanations by Dr. Leopold Stein (Frankfurt: Heinrich Keller, 1886). Also see Norman Kleebat, *Illustrating Jewish Life Styles on Opposite Banks of the Rhine: Alphonse Lévy's Alsatian Peasants and Moritz Daniel Oppenheim's Frankfurt Burghers*, Jewish Art (1990) 16:53–63.

2. Exact meaning of *hadar* uncertain; traditionally understood as "citron."

3. Leviticus 23:39–43.

4. A *mitzvah* (pl. *mitzvot*) is one of 613 prescribed precepts or commandments. These range from celebration of festivals, relevant here, to the prohibition against deviating from traditional authority (Chapter 3), on to the wearing of ritual fringes on a four-cornered garment (Chapter 1).

5. Talmud, Sanhedrin 73a. There was much controversial use of this concept of *rodef* surrounding the assassination in 1995 of Prime Minister Rabin. See the end of Chapter 5, "The Flag That Came out of the Blue."

6. Ephraim Oshry, *Responsa from the Holocaust*, Y. Leiman, trans. (New York: Judaica Press, 1983), pp. 34–35. Our translation is from the more extensive discussion in the five-volume Hebrew edition, published under the title *Sh'eilos uTshovos MiMa'makim*. (Brooklyn: Gross Brothers, 1975), Part 4, Question 6, pp. 49–52. For a collection of responsa from the same era, see Robert Kirsch, *Rabbinic Responsa of the Holocaust Era* (New York: Schocken, 1985).

7. The word "Torah" is used in two senses. Its restrictive meaning refers to the Pentateuch. It can also refer to the entire corpus of Jewish teachings, which speaks always and everywhere, in the sense that God gives the Torah here and now and gives it as well every time a Jew studies Torah, an act considered one of the most essential *mitzvot*. See Jacob Neusner, *Classics of Judaism* (Louisville: Westminster John Knox Press, 1995), p. xv.

8. Good introductions to the oral law and Talmud are found in Adin Steinsaltz, *The Essential Talmud*, Chaia Galai, trans. (New York: Basic Books, 1976); and Jacob Neusner, *Invitation to the Talmud: A Teaching Book* (San Francisco: Harper & Row, 1984).

9. There is an illuminating discussion of the nonbinding nature of *aggadah* by one of the greatest of Jewish religious scholars, Nahmanides. In 1275 he was in the midst of the most tolerant of several Jewish-Christian debates staged in medieval Europe—the Disputation at Barcelona. His opponent was a converted Jew, the Dominican Fray Pul (Brother Paul, or Pablo Christiani). Fray Pul tries to adduce an argument that the Talmud admits that the Messiah has already come. Here is part of Nahmanides's on-stage rejoinder:

> Fray Pul asked me whether the Messiah, of whom the prophets spoke, had already come, and I said that he had not come. He then brought a book of *aggadah* (homilies) in which it is stated that the Messiah was born on the day the Temple was destroyed. I said, even though, that I do not believe in this, that passage would support my words. I shall now explain to you why I said that I do not believe in this [passage]. You should know that we have three kinds of books. The first is the Bible, and all of us believe in it in perfect faith. The second is what is called the Talmud. There are 613 Commandments in the Torah. We [firmly] believe in [the Talmud's] explanation of the commandments. We have a third book called *Midrash,* meaning sermons. It is just as if the bishop

would rise and deliver a sermon, and one the listeners whom the sermon pleased recorded it. With regard to this book [of sermons] if one believes in it, it is well and good; if one does not believe in it, he will not be harmed [spiritually]. We have sages who wrote that the Messiah will not be born until the time near the end [of the exile], at which time he will come to redeem us from exile.

Therefore I do not believe the statement of this book that he was born on the day of the destruction. We also call [the *Midrash*] the book of *Aggadah*, meaning *Razionamiento*. That is to say, it is nothing more than matters which one person tells another.

[Ramban (Nahmanides), *The Disputation at Barcelona*, Charles B. Chavel, trans. (New York: Shilo, 1983).]

Nahmanides was pushed a little by the debate. There is substance in *aggadah* too, as he reasons elsewhere. A reading of the entire text of this remarkable debate is most rewarding. There is also extant a brief Christian account of the debate, predictably forwarding a very different view of the outcome. See Hyam Maccoby, *Judaism on Trial: Jewish-Christian Disputations in the Middle Ages* (Rutherford, N.J.: Farleigh Dickinson University Press, 1982).

10. Collections in English include Abraham Yaakov Finkel, *The Responsa Anthology* (Northvale, N.J.: Jason Aronson, 1990); Nisson E. Shulman, *Authority and Community: Polish Jewry in the Sixteenth Century* (New York: Ktav and Yeshiva University Press, 1986); Alexander Guttmann, *The Struggle over Reform in Rabbinic Literature* (Jerusalem: World Union for Progressive Journalism, 1977); Louis Jacobs, *Theology in the Responsa* (London: Routledge & Kegan Paul, 1975).

11. There are no upper- and lower-case distinctions in Hebrew. The Torah scroll and other ancient manuscripts have almost no punctuation.

12. See Mishnah, Sukkah 1:4 and 1:5 for the full text.

13. Talmud, Sukkah 23a–24a. To a reader objecting to the digressive nature of our writing, we would cite in self-defense this talmudic passage.

14. The elephant does not meet the criteria specified in Leviticus 11 of creatures fit for consumption by observant Jews.

15. Talmud, Sukkah 23a.

16. Exodus 15:2.

17. Jacob Z. Lauterbuch, trans., *Mekhilta de-Rabbi Ishmael* (Philadelphia: JPS, 1933), Volume 2, p. 25. We were led to this *midrash* by Philip Goodman, *The Sukkot and Simhat Torah Anthology* (Philadelphia: Jewish Publication Society, 1973), pp. 224–28. See note 31, Chapter 8, page 346.

18. Rashi on Leviticus 23:43; see Exodus 13:21, "The Lord went before them in a pillar of cloud by day. . . . "

19. Maimonides, *Mishneh Torah*, Book of Seasons, Sukkot 4:16. English translation: Moses Maimonides, *The Code of Maimonides* (New Haven: Yale University Press, 1949).

20. In the land of Israel only the first day is special; but in the diaspora two days have the special status of a festival.

21. The Bar Ilan laser disc and CD-ROM contains not only responsa but also the Talmud and *midrashim*. Bar Ilan Judaic Library, Bar Ilan University (Spring Valley, N.Y.: Torah Educational Software, 1994).

22. Try <ask the rabbi>ohr@virtual.co.il

23. Talmud, Sukkah 30a.

24. Talmud, Sukkah 8a.

25. Talmud, Pesahim 94b.

26. Talmud, Sukkah 22a.

27. Adin Steinsaltz, *op. cit.*, p. 193; S. R. Hirsh, *Horeb* (London: Soneino, 1981), p. 580.

28. Ritual hand ablution is required before a Jew eats bread; that can only be done on already clean hands.

29. For instance, for the status of plastic dishware see J. David Bleich, *Contemporary Halachic Problems*, Vol. II (New York: Ktav, 1983), pp. 43–45.

30. Richard Siegel, Michael Strassfeld, and Sharon Strassfeld, *The Jewish Catalogue* (Philadelphia: The Jewish Publication Society, 1973), pp. 129–30.

31. *The Jewish Catalogue, ibid.*, p. 129.

32. Joseph Caro, *Shulhan Arukh, Orach Hayim* 631, Part 3, Gloss 6 in *Mishnah Brurah*.

33. For a charming story of how the U.S. Army Engineers dealt with the construction of a *sukkah* in postwar Germany, see Oscar M. Lifshutz, "A Sukkah by Courtesy of the U.S. Army Engineers," in Philip Goodman, *op. cit.*

Chapter 3: You Must Not Deviate to the Right or the Left

1. Stephen Jay Gould ["Double Entendre," The Sciences (July/August 1995): 36] made a similar point:

> Most of us read the creation story of Genesis I as a tale of addition—first God creates the earth, then plants, fishes, terrestrial beasts and finally exalted us. But more literal attention to the words strongly suggests differentiation as the intended theme. From an initial formless chaos, God makes a series of progressively finer separations: earth from firmament, light from darkness, land

from sea, coalescence of sun and moon as sources of light, "bringing forth" of living things from the earth.

2. Roald Hoffmann, *The Same and Not the Same* (New York: Columbia University Press, 1995).

3. *"Raffiniert ist der Herrgott, aber boshaft ist Er nicht,"* as cited by Ronald W. Clark, *Einstein, An Illustrated Biography* (New York: Harry N. Abrams, 1984), p. 232.

4. Harry B. Gray, in his review of F. A. Cotton, "Chemical Applications of Group Theory," *Journal of Chemical Education* (1964) 41(2): 113–14. This amusing document was brought to our attention by Derek A. Davenport, whom we thank.

5. Avigail Sheffer and Hero Granger-Taylor, "Textiles from Masada," in *Masada IV, The Yigal Yadin Excavation 1963–1965* (Jerusalem: Israel Exploration Society, 1994), p. 162.

6. The hardly unfamiliar hand is, of course, from Michelangelo's fresco of the Creation on the ceiling of the Sistine Chapel.

7. G. Ohloff, *Scent and Fragrances* (Berlin: Springer, 1990), pp. 4–45; Edwin Thall, "When Drug Molecules Look in the Mirror," *Journal of Chemical Education* (1996) 73: 481–84; Stephen C. Stimson, "Chiral Drugs," *Chemical and Engineering News* (October 9, 1995): 44–72. See also Malcolm W. Browne, "Mirror Image Chemistry Yielding New Products," *New York Times* (August 13, 1991): C1, C8. The left-handed sugar story comes from this source. See also the amusing tale of David Hall, "Coming to Grips with Left-Handed Yoghurt," *New Scientist* (7 September, 1991): 62.

8. Roald Hoffmann, "Specula," unpublished.

9. So we call them. But remember the arbitrariness of language—what matters is not what we call left (or *sinister,* or *smol,* or *links*) and right (or *droite,* or *yamin,* or *rechts*), but that they are different. It happens that the naturally occurring amino acids have their atoms arranged in one of two ways, a way that corresponds to the arrangement in a reference molecule that happens to rotate the plane of polarized light to the left. But more on that later.

10. This figure also appears in a wonderful book by István and Magdolna Hargittai, *Symmetry, A Unifying Concept* (Bolina: Shelter, 1994), p. 32.

11. For a perceptive and entertaining discussion of many of the subjects of this chapter, see Martin Gardner, *The New Ambidextrous Universe*, 3rd rev. ed. (New York: W. H. Freeman). A readable introduction to chirality is Roger A. Hegstrom and Dilip K. Kondepudi, "The Handedness of the Universe," *Scientific American* (January 1990): 108–15.

12. Rashi on Deuteronomy 17:11, citing Sifrei.

13. It is very well told in Jean Jacques, *The Molecule and Its Double*, Lee Scanlon, trans. (New York: McGraw-Hill, 1993). The original French version, under the title *La Molecule et son Double,* was published by Hachette in 1992.

14. It's not easy, but with some effort it's possible for students to reproduce Pasteur's classic experiment. This we owe to George B. Kauffman and Robin D. Myers, "The Resolution of Racemic Acid," *Journal of Chemical Education* (1975) 52: 777–81.

15. Louis Pasteur, *Researches on the Molecular Asymmetry of Natural Organic Products* (Edinburgh: The Alembic Club, 1915), pp. 20–21. These lectures were given by Pasteur in 1860.

16. For a crystal clear, balanced account of the background of Pasteur's discovery, see Seymour H. Mauskopf, "Crystals and Compounds: Molecular Structure and Composition in Nineteenth-Century French Science," *Transactions of the American Philosophical Society* (1976) 66(3): 5–82. An attempt at a revisionist account is given by Gerald L. Geison, *The Private Science of Louis Pasteur* (Princeton: Princeton University Press, 1995), Chapter 3.

17. We were reminded of this by Alex Altman, letter to Roald Hoffman, December 5, 1995. There is an intriguing connection between Rashi's livelihood and his religious work: see Haym Soloveitchik, "Can Halakhic Texts Talk History?" *American Jewish Studies Review* (1976) 1: 153–96.

18. Jerusalem Talmud, Horayot 1:1.

19. Nahmanides, *Ramban: Commentary on Deuteronomy,* Charles Chavel, trans. (New York: Shilo, 1976), p. 207, on Deuteronomy 17:11.

20. There are several titles that are variations on "rabbi," the root word referring to "great" (in learning)—for example, *rabban* (Aramaic), *rabbenu* (our rabbi), *rav* (Babylonian talmudic scholar), and *rebbe* (usually hasidic).

21. Mishnah, Rosh Hashanah 2:8–10.

22. Cyril Domb (letter to Roald Hoffmann, February 11, 1996) remarks that our interpretation of this encounter is not completely accurate. He says:

> In that period the new month was fixed by experimental observation, and witnesses who claimed to have seen the new lunar crescent would be subjected to detailed cross-examination. The issue between the two was whether to accept the evidence of the witnesses part of whose testimony was patently false. Rabban Gamaliel accepted the part of the evidence which suited him (it is very likely that he had already had a calculated value) whereas R. Yehoshua maintained that once [the witnesses] had been established as unreliable none of their evidence could be accepted.

23. Much of the material we discuss in this chapter is drawn from and treated in a much deeper, more scholarly way, in a special issue of the

journal *Tradition* dealing with rabbinic authority [Volume 27, Number 4 (Summer 1993)].

24. There is a fascinating story, which we can only sketch here, related to the ability of organisms (really their molecules) to attack selectively one mirror image and not another. There is a natural hormone, vasopressin, useful in the treatment of diabetes as an antidiuretic. It's a cyclic peptide, a ring made up of several amino acids. The problem with vasopressin as a drug is that it just doesn't survive very long in the body. Why? Because, quite naturally, some of our own enzymes get at it and break it up.

So people thought up the following ingenious strategy: Make the vasopressin a little bit unnatural, by replacing one, and just one, of its amino acids by its mirror image. The replacement is exquisitely chosen not to affect the therapeutic action. But it does throw a mirror-image wrench into the breakdown mechanisms, slowing them down. The result—a longer-lasting antidiuretic drug, DDAVP.

For more on the history of this inspired piece of molecular engineering, see V. Pliska and H. Vilhardt, "Design of DDAVP. The Path Between Ideas and Pharmocologic Reality," in A. H. Sutor, *Minirin, DDAVP Anwendung bei Blutern* (Stuttgart: Schattauer, 1980), pp. 12–29. We are grateful to Dr. Edward Wolpow for bringing this story to our attention.

25. See Stimson and Browne in note 7.

26. The important paper, J. M. Bijvoet, A. F. Peerdeman, and A. J. Van Bommel, "Determination of the Absolute Configuration of Optically Active Compounds by Means of X-Rays," *Nature* (August 18, 1951): 271–72, was preceded by another theoretical contribution by the same group: A. F. Peerdeman, A. J. Van Bommel, and J. M. Bijvoet, "Determination of Absolute Configuration of Optical Active Compounds by Means of X-Rays," *Proceedings of the Section of Sciences of the Koninklijke Nederlandse Akademie Van Wetenschappen* (1951) 54(B): 16–19.

27. Stephen Weininger, letter to Roald Hoffmann, April 14, 1996.

28. Alfred Bader, *The Bible Through Dutch Eyes* (Milwaukee: Milwaukee Art Center, 1976), p. 7.

29. Wolfgang Stechow, "*Jacob Blessing the Sons of Joseph* by Adriaen van der Werff," *Allen Memorial Art Museum Bulletin* (1965) XXII(2): 69–73; Wolfgang Stechow, "*Jacob Blessing the Sons of Joseph,* from Rembrandt to Cornelius," in Antje Kosegarten and Peter Tigler, *Festschrift Ulrich Middledorf* (Berlin: Walter de Gruyter, 1968), pp. 460–65. For older portrayals of this subject, see Walter Stechow, "*Jacob Blessing the Sons of Joseph;* From Early Christian Times to Rembrandt," *Gazette des beaux-arts* (1943) XXIII: 193–208.

30. Heinrich Wölfflin. "Uber das Rechts und Links im Bilde," in *Gedanken zur Kunstgeschichte* (Basel: Benno Schwabe, 1940), pp. 82–96.

31. Richard D. Buck, "Adriaen van der Werff and Prussian Blue," *Allen Memorial Art Museum Bulletin* (1965) XXII(2): 71–76.

32. A translation of these epochal papers of van't Hoff and Le Bel is given by O. Theodor Benfey, ed., *Classics in the Theory of Chemical Combination* (New York: Dover, 1963). See also the various contributions in O. Bernard Ramsey, ed., *van't Hoff–Le Bel Centennial* (Washington, D.C.: American Chemical Society, 1975).

33. For the intellectual setting here, see the perceptive and comprehensive biography of Kolbe by Alan J. Rocke, *The Quiet Revolution: Herrmann Kolbe and the Science of Organic Chemistry* (Berkeley: University of California Press, 1993).

34. G. W. Wheland, *Advanced Organic Chemistry*, 3rd ed. (New York: John Wiley, 1960), pp. 197–98.

35. Rocke's book, *op. cit.* (pp. 329–30), contains excerpts of Wislicenus and van't Hoff's replies to Kolbe's tirade. Wislicenus writes to Kolbe:

> You cannot have possibly studied van't Hoff's essay . . . [for] how else could you have reproached me (by logic I do not understand) for a tendency toward spiritualism, or held against the young van't Hoff his position at a veterinary school, or against the translator Herrmann, who was my assistant and solely due to pressing external circumstances accepted a position at the agricultural institute in Heidelberg! I have never doubted that it is a holy zeal for the truth that guides your critical pen; but on the other hand I regret that you do not seem to concede any possibility of your own fallibility, which everyone must grant. . . . I know that I can err, but I also know that I have no cause to allow myself to be struck from the ranks of exact scientists, for I as well as you have the will to serve the truth. . . .

And van't Hoff, in the pages of the leading chemical journal of its time:

> A theory that so far is contradicted by no single fact can only be further examined experimentally. Thus when someone, even so fine a chemist as Kolbe, avers that a chemist who is not yet well known and who is employed at a veterinary school should not bother himself with theories . . . I can only say that such behavior fortunately is not a sign of the times, but rather must be regarded as a contribution to understanding a single individual.

36. Mirrors are important, too; for a sexy context see Midrash Tanhuma on Exodus 38:8, as discussed by Nehama Leibowitz, *Studies in Shemot (Exodus)* (Jerusalem: World Zionist Organization, 1981), pp. 689–95.

37. *Halitzah* (literally, "removal of a shoe") is the ceremony whereby a childless widow is released from the biblical obligation to marry her late husband's brother. In the ceremony the widow removes the brother's right shoe.

We are grateful to Rabbi Eli Silberstein for bringing to our attention a fascinating responsum on fixed-versus-relative interpretation of left-handedness by Moshe Sofer (Hatam Sofer, d. 1839).

38. The painting by Marc Chagall at left is titled *Jew in Black and White*, 1914, oil on canvas (104× 84 cm), Collection Stiftung Karl und Jürg Im Obersteg. Other versions are in the Museo d'Arte Moderna, Venice and the Art Institute of Chicago. The etching at the right is titled *The Grandfathers*. It first appeared in Marc Chagall, *Mein Leben* (Berlin: Paul Cassirer, 1923). 110 copies were printed. The one reproduced here is in the Sprengel Museum, Hanover.

39. The complexity of this issue is reflected in the 125-page compendium by Abraham Halevi Horwitz, *Book of Halakhah, Part II, Laws of Tefillin for the Left-Handed and Ambidextrous* [Hebrew] (Bene Brak: D. BenNun, 1978). See also Eliyahu Elharat, *Laws of Left-Handedness* (Jerusalem: Ketavim, 1986) and Daniel Eidensohn, *Yad Israel: Comprehensive Subject Index to Responsa of R. Moshe Feinstein* (Jerusalem: Daniel Eidensohn, 1994), which has 40 references.

40. For references to not one, but several fascinating literatures we are unable to delve into, see: (a) Stanley Coren, *The Left-Hand Syndrome* (New York: Free Press, 1992); Sally P. Springer and George Deutsch, *Left Brain, Right Brain* (New York: W. H. Freeman, 1993); (b) G. E. R. Lloyd, "Right and Left in Greek Philosophy," in *Methods and Problems in Greek Science* (Cambridge: Cambridge University Press, 1991), pp. 27–48; and Rodney Needham, ed., *Right & Left: Essays on Dual Symbolic Classification* (Chicago: University of Chicago Press, 1973).

41. Avraham Yaakov Finkel, *The Responsa Anthology* (Northvale, N.J.: Jason Aronson, 1990), pp. 176–77.

42. From "Left Right Out of Your Heart" ("Hi-Lee Hi-Lo Hi-Lup-Up-Up"), words by Earl Shuman, music by Mort Garson. Copyright 1958 by Shapiro, Bernstein and Co.

43. The twentieth one, also the simplest one—glycine—is not chiral!

44. An excellent account of the origins of molecular assymmetry may be found in Stephen F. Mason, "From Pasteur to Parity Violation: Cosmic Dissymmetry and the Origins of Biomolecular Handedness," *Ambix* (1991) 38(2): 85–99. See also Vladik A. Averisov, Vitalii I. Goldanskii, and Vladimir V. Kuz'min, "Handedness, Origin of Life and Evolution," *Physics Today* (July 1991): 33–41.

45. François Jacob, "Evolution and Tinkering," *Science* (1977) 196: 1161–66.

46. Quoted by Jon Cohen, "Getting All Turned Around Over the Origins of Life on Earth," *Science* (3 March 1995) 267: 1265–66.

47. See Stephen F. Mason, *op. cit.*

48. Talmud, Shabbat 21b–22a.

49. There is little question that one should debate. Here for instance is an account (Baba Metzia 84a) of an episode in the life of Rabbi Yohanan. His friend Resh Lakish dies (the tragic circumstances are intriguing, but not relevant here). To console Yohanan his rabbinical colleagues send one of their own to study with him, Rabbi Elazar Ben Padat, who, in an attempt to be nice to Yohanan, after every statement by the great rabbi says: "Yes. Quite. There's a law that supports you." Then Rabbi Yohanan cries out:

> Are you the son of Lakish? When I would say something, the son of Lakish would raise twenty-four objections against me, and I would give him twenty-four answers, and the statement would thereby be clarified. And you say: "There's a law that supports you."

50. Talmud, Baba Metzia 59b.

51. *Ibid.*

52. Rabbi Yehoshua quotes from Deuteronomy 30:12, part of a moving speech of Moses to the Israelites:

> Surely, this Instruction which I enjoin upon you this day is not too baffling for you, nor is it beyond reach. It is not in the heavens, that you should say, "Who among us can go up to the heavens and get it for us and impart it to us, that we may observe it?" Neither is it beyond the sea, that you should say, "Who among us can cross to the other side of the sea and get it for us and impart it to us, that we may observe it?" No, the thing is very close to you, in your mouth and in your heart, to observe it.

53. Attributed to R. Akiva, Talmud, Baba Kamma 41b.

54. *Sheelot u-Teshuvot ha-Rosh* LV:9, as quoted by Aaron Kirschenbaum, "Mara de-Atra: A Brief Sketch," *Tradition* (1993) 27(4): 35–40; Shnayer Z. Leiman, in "Dwarfs on the Shoulders of Giants," *ibid.*, pp. 90–94, points out that while the earlier rabbis were giants and the later generations were dwarfs, dwarfs see farther (if they sit on the shoulders of giants). A propos of which, see Robert K. Merton, *On the Shoulders of Giants: A Shandean Postscript* (New York: Harcourt, Brace & World, 1965).

55. Nahmanides, Introduction to *Milhamot Hashem*, as cited by Jeffrey R. Woolf, "The Parameters of Precedent in *Pesak Halakhah*," *Tradition* (1993) 27(4): 41–48.

56. Talmud, Baba Batra 130b–131a, as cited by Jeffrey R. Woolf, note 55.

57. For an excellent general discussion of authority, see Norman Lamm and Aaron Kirschenbaum, "Freedom and Constraint in the Jewish Judicial Process," *Cardozo Law Review* (1979) 1: 99–133.

58. Observant Israeli men are not exempt from the universal draft, but any man engaged in full-time study of Torah may receive a renewable, long-term deferment. Many orthodox choose the option of a *yeshivat hesder*, which combines religious study with alternating periods of military service.

59. The following rulings relate to public policy and as such differ greatly from halakhic rulings on ritual and personal questions. "Daas Torah [An *Halakhic* Opinion] Issued by Prominent Rabbis in Eretz Yisroel Concerning Territorial Compromise," *Jerusalem Post* (July 20, 1996); Shlomo Chesna, Menachem Rahat, and Baruch Meiri, "Rabbis Renew *Halakhic* Ruling to Soldiers: Do Not Evacuate Hebron" [Hebrew], *Maariv* (December 26, 1996); Amnon Bazak, You Shall Live by These (Jerusalem: Meimad, 1995). See also J. D. Bleich, sections on "The Sanctity of the Occupied Territories," and "Judaea and Samaria: Settlement and Return," in *Contemporary Halakhic Problems, Volume II* (New York: Ktav, 1983). There is also relevant material in Volume III of this more useful work.

60. For some interesting reflections on Rabbi Abraham Isaac Kook's ideas on limitations to freedom of thought in Jewish tradition, see Shalom Carmy, "Dialectic, Doubters, and a Self-Erasing Letter: Rav and the Ethics of Belief," in *Rabbi Abraham Isaac Kook and Jewish Spirituality*, Lawrence J. Kaplan and David Shatz, eds. (New York: New York University Press, 1995), pp. 205–36.

61. Sam Orbaum, "The National Census & the Jewish Question," *The Jerusalem Post* (October 27, 1995): 31.

62. See also Gerard E. Dallal and Aviva Must, "Dreidel Rotation: Statistical Evidence That God Is Left-Handed," *Journal of Irreproducible Results* (1992) 37: 26–27.

63. Robert E. Kohler, Jr., "Irving Langmuir and the 'Octet' Theory of Valence," in *Historical Studies in the Physical Sciences*, Russell McCormmach, ed. (Princeton: Princeton University Press, 1975), pp. 73, 77, 84. We thank Derek A. Davenport for bringing these quotations to our attention via his hilarious and thought-provoking lecture, "Opprobium: Occurrence, Preparation, Properties, and Uses."

64. The physicist daughter of one of us [R. H.] Ingrid H. H. Zabel, comments: "But I think most physicists would enjoy citing a paper of Einstein's from the '20s or Rayleigh's from the last century."

65. For a wise analysis of peer review past and future, see David Goodstein, "Peer Review After the Big Crunch," *American Scientist* (1995) 83: 401–02.

66. Roald Hoffmann, "Under the Surface of the Chemical Article," *Angewandte Chemie, International Edition English* (1988) 27: 1593.

67. Roald Hoffmann, Vladimir I. Minkin, and Barry Carpenter, "Ockham's Razor in Chemistry," *Bulletin de la Societé Chimique de France* (1996) 133: 117–30.

68. For a wonderful account of the many ways in which one great scientist claimed authority (and enraged others in the process), see Maurice Crossland, "Lavoiser, the Two French Revolutions, and the Imperial Despotism of Oxygen," *Ambix* (1995) 42(2): 101–18.

69. The parallels in rabbinical and scientific reasoning are discussed in an intriguing way by Menachem Fisch, "The Perpetual Convenant of Jewish Learning," in *Summoning: Ideas of the Covenant and Interpretative Theory*, Ellen Spolsky, ed. (Albany: State University Press of New York, 1993), pp. 91–114.

70. Michael S. Berger, "Rabbinic Authority: A Philosophical Analysis," *Tradition* (1993) 27(4): 61–71.

71. Rabbi Shalom Carmy, letter to Roald Hoffmann, September 19, 1995. The wonderful possibility outlined is reminiscent of the double-slit experiment/paradox in quantum theory.

72. Israel L. Shenker, "Responsa: The Law as Seen by Rabbis for 1,000 Years," *New York Times* (May 5, 1975): 33. Rabbi Feinstein's obituary appears in the March 15, 1986, issue of the *New York Times*.

73. Cyril Domb, letter to Roald Hoffmann, February 11, 1996.

74. See Shalom Carmy, note 71.

75. Edmund Gittenberger, "On the One Hand. . . . ," *Nature* (5 January 1995) 373: 19.

76. And in heraldry, right is called left and left right, the justification being that it is the perpective of the shield-bearer that matters. See this used in a fascinating context by Michel Tournier, *The Ogre*, Barbara Bray, trans. (Garden City: Doubleday, 1972), pp. 303–04.

77. Jonah 4:11.

Chapter 4: Bitter Waters Run Sweet

1. For most of the biblical passages, the translation by the Jewish Publication Society (JPS) was used: *The Holy Scriptures* (Philadelphia: Jewish Publication Society, 1962). The authors, however, have retranslated the important passage in this chapter, Exodus 15: 22–25. The key word in this passage is *etz*, which can mean "tree," or "wood," or "log." We translate it as "tree."

Some of the messages in this chapter are real e-mail messages, which we have reproduced with the permission on the authors. We have left those letters in their original style, including the spelling of the name of the Deity, capitalization, and punctuation, which may differ from the conventions followed in the rest of the book. Some people (participants in the e-mail exchanges, authors of works cited) prefer to follow the Hebrew usage of not writing the Deity's name even in English, and so modify it to G-d.

2. Exodus 15:22–25.

3. Abraham ibn Ezra (b. Toledo 1092), commentary on Exodus 15:25, *Torat Chaim Chumash, Shemot,* Vol. 1 [Hebrew] (Jerusalem: Mossad Harav Kook, 1987), p. 195.

4. Quoted, with permission, from an e-mail message by Andy Goldfinger.

5 A *midrash* (Talmud, Yoma 65b) points out that manna was bland and tasteless but assumed the taste of what your heart, or stomach, desired. So if one Israelite yearned for southern fried chicken and another had a yen for chocolate mousse, then the manna tasted just like those delicacies to the individuals concerned.

6. "Yekke" is a mildly derogatory term applied to Jews of German origin by other Jews. It has a sense of "those people taking on airs," or acting "better than thou," based on their origins in what was supposed to be the high culture of its day.

7. *Midrash Tanhuma* [Hebrew], Hanoch Zundel, ed. (Jerusalem: Lewin-Epstein Bros.,1952) B'shalah, XXII, p. 191. This is a *midrash* on the weekly Torah portions, attributed to Rabbi Tanhuma B. Abba. The date of composition of this *midrash* is debated; it lies between 400 and 900 C.E. This collection was also called "Yelamdeinu," because each section begins with this Hebrew word, whose sense is roughly "Let our masters teach us. . . ."

8. *Midrash Mekhilta.* Attributed to Rabbi Ishmael, this is a *midrash* of laws and exegesis on the Book of Exodus. See also Menahem M. Kasher, *Encyclopedia of Biblical Interpretation* (New York: American Biblical Encyclopedia Society, 1953–79), Vol. VIII, p. 203.

9. Nogah Hareuveni, *Desert and Shepherd in Our Biblical Heritage* (Neot Kedumim: Biblical Landscape Reserve in Israel, 1991), pp. 56–57.

10. Yehuda Felix, the expert on flora and fauna in Jewish tradition, states that the most likely candidate of the *midrash* is the oleander: Yehuda Felix, *Trees in the Bible and Talmudic Literature,* Vol. 2 [Hebrew: *Atzey bsamim, yaar, v'noy*] (Jerusalem: Rubin Mass Press, 1997), see entry for *Harduf* (oleander), p. 179.

11. For more on olive processing, why Greek olives are strong, but California ones mild, see Harold McGee, *On Food and Cooking* (New York: Collier, 1988), p. 204.

12. Chizkuni, thirteenth-century commentary of Hezekiah ben Manoah. See *Torat Chaim Chumash, Shemot,* Vol. 1 [Hebrew] (Jerusalem: Mossad Harav Kook, 1987), p. 194.

13. See Yair Zakovich, *The Concept of the Miracle in the Bible* (Tel Aviv: Ministry of Defense Books, 1991); and the Summer 1986 issue of *Da'at* (*Journal of Jewish Philosophy and Kabbalah*) No. 17, published by Bar Ilan University Press.

14. George Bernard Shaw, in the Preface to *The Simpleton of the Unexpected Isles, The Six of Calais & The Millionairess* (New York: Dodd, Mead, and Co., 1936), pp. 4–5.

15. Yisrael Meir Hacohen (The Hafetz Hayim), *Mishnah Berurah, Shulhan Aruch Orach Hayim,* Vol. II(C), Aviel Orenstein, ed. (Jerusalem: Pisgah Foundation, 1989), p. 267.

16. J. David Bleich, *Contemporary Halakhic Problems,* Vol. 2 (New York: Ktav/Yeshiva University Press, 1983), pp. 222–23.

17. For further information, see Kenneth L. Woodward, *Making Saints: How the Catholic Church Determines Who Becomes a Saint, Who Doesn't, and Why* (New York: Simon and Schuster, 1990); also the thoughtful column by Peter Steinfels, "Beliefs: Will It Take a Miracle to Save Miracles from the Toy Department of Modern Life?" *The New York Times* (December 16, 1995): 13; Nancy Gibbs, "The Message of Miracles," *Time* (April 10, 1995): 42–51.

18. Yeshayahu Leibowitz, *Notes and Remarks on the Weekly Parashah,* Shmuel Himelstein, trans. (Brooklyn: Chemed Books, 1990), p. 70.

19. Yeshayahu Leibowitz, *ibid.,* p. 71.

20. Augustine, *De utilitate credende,* 16.34. This reference comes from a most readable recent article by Peter Harrison, "Newtonian Science, Miracles, and the Laws of Nature," *Journal of the History of Ideas* (1995) 56(4): 531–53.

21. The reference is to the aforecited *Midrash Tanhuma.*

22. Nahmanides (Ramban), *Commentary on the Torah, Exodus,* Charles B. Chavel, trans. (New York: Shilo, 1973), p. 211. The distinction between the meanings of *vayoreihu* and *vayar'eihu* was pointed out in a *midrash* by R. Shimon bar Yohai. See Kasher, *op. cit.* Nahmanides applies this in an innovative way.

23. Russell L. Rouseff, ed., *Bitterness in Food and Beverages* (Amsterdam: Elsevier, 1990).

24. "Rotting Straw Stops Algal Blooms," *Chemistry in Britain* (1991) 28: 607.

25. Apparently an excessive burden of certain algae can lead to bitter tastes: See Frederick W. Pontius, ed., *Water Quality and Treatment: A Handbook of Community Water Supplies* (New York: McGraw-Hill, 1980), pp. 150–51. See also Robert C. Hoehn, "Biological Causes of Tastes and

Odors in Drinking Water Supplies," *Water Quality Bulletin* (1988) 13(2, 3): 46–51. We are grateful to Herbert Deinert and Mark R. Deinert for bringing these references to our attention.

26. For a discussion of water supplies in the Talmud, see Samuel S. Kottek, "Gems from the Talmud: Public Health I—Water Supply," *Israel Journal of Medical Science* (1995) 31: 255–56.

27 S. Fletcher, "Algicidal Activity," *Chemistry in Britain* (1991) 24: 1114.

28. Paul Day, "Chemical Explanation of Marah," *Chemistry in Britain* (1992) 21: 227.

29. Norman Goldberg, quoted with permission.

30. Fredric M. Menger, quoted with permission.

31. Entry for *Marah, Encyclopedia shel Hatanakh* [Hebrew] (Jerusalem: Jerusalem Publishing House and Yediot Aharonot, 1987).

32. Menahem M. Kasher, *Encyclopedia of Biblical Interpretation, Exodus,* Vol. VIII, Harry Freedman, trans. (New York: American Biblical Encyclopedia Society, 1953–79), p. 165, footnotes 255–256. He cites a *midrash* that specifies Bir (Ain) Hawarrah; he also mentions the Arab tradition that the place was a bitter spring today called Ain Mussah (or Uyun Musa), "spring of Moses."

33. *Midrash Tanhuma Yashan, Vayakhel* 9: "Rabbi Levi said, they were bitter, the generation was bitter in its deeds." Another *midrash* attributes this sentiment to R. Yehoshua: Menahem M. Kasher, *op. cit.,* p. 202. It is interesting in this context to note that Proverbs 27:7 says "A sated person disdains honey but to a hungry man bitter seems sweet."

34. G. M. Friedman and W. E. Krumbein, eds., *Hypersaline Ecosystems: The Gavish Sabkha* (Berlin: Springer, 1985).

35. Wolfgang E. Krumbein, quoted with permission.

36. Arie S. Issar, *Water Shall Flow from the Rock: Hydrogeology and Climate in the Lands of the Bible* (Berlin: Springer, 1990), p. 109–22.

37. Nehama Leibowitz, *Gilyonot B'Farshat Hashavua* (5714), *B'shalah* (Jerusalem: Education Ministry, 1954).

38. Talmud, Pesachim 36a, in the discussion of the *marror*, bitter herbs, of the Passover Seder.

39. *Midrash Mekhilta,* quoted in Kasher, *op. cit.,* p. 204.

40. See Rouseff, *op. cit.*

41. Sweetness is the taste best explored—it is easier, among other things, to get people to taste sweet things by way of experimentation than bitter ones. There's also a market catering to the sweet tooth in us all, and good medical reasons for some people to avoid sugar.

For an introduction to the chemistry of sweetness, see Jerry W. Ellis, "Overview of Sweeteners," *Journal of Chemical Education* (1995) 72: 671–75; A. Douglas Kinghorn and Edward J. Kennelly, "Discovery of Highly Sweet Compounds from Natural Sources," *Journal of Chemical Education* (1995) 72: 676–80; D. Eric Walters; "Using Models to Understand and Design Sweeteners," *Journal of Chemical Education* (1995) 72: 680–83.

42. Diane Ackerman, *A Natural History of the Senses* (New York: Random House, 1990), p. 139.

43. For references to the extensive work on *Gymnema* taste modifiers, see Bruce P. Halpern, "The Use of Vertebrate Laboratory Animals in Research on Taste" in William I. Gay, ed., *Methods of Animal Experimentation* Vol. IV (New York: Academic Press, 1973), pp. 225–362.

44. See Linda M. Kennedy and Bruce P. Halpern, "Extraction, Purification and Characterization of a Sweetness-Modifying Component from Zizyphus Jujuba," *Chemical Senses* (1980) 5: 123–47, and references therein.

45. Thomas Eisner and Bruce P. Halpern, "Taste Distortion and Plant Palatability," *Science* (25 June 1971) 172: 1362.

46. Kazuko Yoshikawa, Naomi Shimono, and Shigenobu Arihara, "Antisweet Substances, J*ujubasaponins I–III* from *Zizyphus jujuba;* Revised Structure of *Ziziphin*," *Tetrahedron Letters* (1991) 32: 7059–62.

47. See the report by A. Maureen Rouhi, "Researchers Unlocking Potential of Diverse, Widely Distributed Saponins," *Chemical and Engineering News* (September 11, 1995): 29–35.
Inter alia:

> Beer lovers stranded in the Amazon jungle of Brazil need not worry about the supply of their favorite drink, thanks to saponins. With a knife, water, and the bark of *Ampdozizyphus amazonica,* anyone can make beer there. Used medicinally as a treatment for the common cold, the bark also provides an instant drink very much like commerical beer. . . . Making beer in the jungle is easy . . . freshly collected bark, which is extremely bitter, is washed with water. The outer layer is shaved off with a knife, and the shavings are soaked in water. The resulting extract is brown and bitter. . . . Before the beer is poured, the liquid is shaken well to form a layer of foam. It tastes like commercial beer but without the alcohol. You can drink as much as you like without getting intoxicated.

48. L. M. Bartoshuk, C.-H. Lee, and R. Scarpellino, "Sweet Taste of Water Induced by Artichoke (*Cynara scolymus*)," *Science* (1972) 178: 988–90.

49. R. J. Kurtz and W. D. Fuller, "Ingestibles Containing Substantially Tasteless Sweetness Inhibitors as Bitter Taste Reducers or Substantially Tasteless Bitter Inhibitors as Sweet Taste Reducers," U.S. Patent 5,232,735, assigned to BioResearch Inc., Farmingdale, N.Y. Also Robert J. Kurtz and William D. Fuller, "Development of a Low Sodium Salt: A Model for Bitterness Inhibition," to be published. This work was kindly brought to our attention by Fred Naider.

50. Numbers 5:17–24.

51. This engraving by Cornelius Nicolaus Schurtz precedes the frontispiece of Johann Christoph Wagenseil's Latin translation of the *Mishnah of Tractate Sotah* (Altdorf, 1674). The representation and its relationship to Andrea del Sarto's *The Cure of the Possessed Woman* is beautifully analyzed by Rachel Wischnitzer, *From Dura to Rembrandt: Studies in the History of Art* (Milwaukee: Aldrich, 1990), pp. 155–58.

52. Zohar 3, 124b, TS 15, 274, as cited by Menahem M. Kasher, *Encyclopedia of Biblical Interpretation* (New York: American Biblical Encyclopedia Society, 1953–79), Vol. VIII, p. 206.

53. Still another bittersweet passage is to be found in the *midrash* on the Song of Songs. The relevant passage is 2:11: "Like an apple tree among trees of the forest, So is my beloved among the youths. I delight to sit in his shade, And his fruit is sweet to my mouth." In the *midrash*, R. Isaac says this passage refers to the twelve months that Israel spent in front of Mt. Sinai regaling themselves with the words of the Torah: "To my taste it was sweet, but to the taste of the other nations it was bitter like wormwood."

54. Quoted, with permission, from a message from Professor Moshe Greenberg.

55. The *wadi* is empty for now. . . . But there is so much more to say, so much more that has been perceptively said. See, for instance, the articles in note 13; and Michael Berger, "Miracles and the Natural Order in Nahmanides," in *Rabbi Moses Nahmanides, Ramban Explorations*, Isadore Twersky, ed. (Cambridge: Harvard University Press, 1983).

56. Exodus 15:25–26.

57. God rebukes Moses in a *midrash* in the Talmud, Shabbat 89a, which concludes, "You should at any rate have given Me a helping hand." Nehama Leibowitz discusses the symbolic cooperation between God and man, and the human assistance that God sometimes requires for man's sake. See Nehama Leibowitz, "The Priestly Blessing," *Studies in Bamidbar (Numbers)* (Jerusalem: World Zionist Organization, 1995), pp. 60–67 and *Studies in Devarim (Deuteronomy)* (Jerusalem: World Zionist Organization, 1980).

Chapter 5: The Flag That Came out of the Blue

1. Talmud, Berakhot 9b.

2. Numbers 8:6, 7.

3. *Midrash* based on Proverbs 14:1.

4. Midrash Numbers Rabbah tells the following story (see also Talmud, Baba Batra 74a):

> Rabba son of Bar Hana related: "I was once walking on the way, when a certain Arab merchant said to me: 'Come, and I will show you the spot where the men of Korah were swallowed up.' I went and saw two cracks from which smoke was coming. The Arab took a ball of clipped wool and, steeping it in water, set it upon the top of a spear and inserted it into that place. The wool was singed and dropped off. He said to me: 'Listen, if you can hear something.' I heard them saying: 'Moses and his Torah are true, and they (Korah and his associates) are liars.' He said to me: 'Every thirty days Gehenna whirls them back to this spot like meat in a pot, and they thus exclaim: "Moses and his Torah are true." In time to come the Holy One, blessed be He, will take them out.'"

[Midrash Rabbah, *Numbers,* Vol. II, H. Freedman and Maurice Simon, eds.; Judah J. Slotki, trans. (London: Soncino, 1983).]

5. An interesting account of ultramarine, the introduction of synthetic ultramarine in the 1820s, and the influence of the availability of the synthetic pigment on Impressionist painting is to be found in Anthony Butler and Rosslyn Gash, "A Higher Shade of Blue," *Chemistry in Britain* (1993) 29: 952–54.

6. Pliny the Elder, *Natural History,* Book IX, H. Rackham, trans. (Cambridge: Harvard University Press, 1940), p. 255.

7. The information in this section derives from Jane Bridgeman, "Purple Dyes in Late Antiquity and Byzantium," in *The Royal Purple and the Biblical Blue: The Study of Chief Rabbi Isaac Herzog and Recent Scientific Contributions,* Ehud Spanier, ed. (Jerusalem: Keter, 1987), pp. 159–65.

8. *Encyclopaedia Judaica* (Jerusalem: Keter, 1973), entry for Korah.

9. A sacred pigment is made from snails whose flesh should not be consumed. Recall the case of elephants which serve as permissible walls for a *sukkah,* yet could not be eaten (see Chapter 2).

10. A group of young people in Israel (*Amutat P'til Tekhelet*: ari@tekhelet.co.il or http://virtual.co.il/orgs/tekhelet/news.htm) has set up an enterprise for producing *tekhelet*. They dive for the murex, and also pay fishermen and divers to collect them. From 1 kilogram of snails they extract approximately 10 grams of snail glands. That in turn produces enough dye to color about two sets of strings for the fringes of the ritual garment.

One partner in this commendable business said "Who would have ever thought that the chemistry we so reluctantly took in college would be of any use?" See Plate 10.

11. See *The Royal Purple and the Biblical Blue*, Ehud Spanier, ed., *op. cit.*; Matthias Seefelder, *Indigo in Culture, Science and Technology*, 2nd ed. (Landsberg: Ecomed, 1994); Menahem Borstein, *HaTekhelet* (Hebrew) (Jerusalem: Sifriyati, 1987); Gösta Sandberg, *Indigo Textiles: Technique and History* (London, A & C Black, 1989); Baruch Sterman, "The Science of Tekhelet," in *Tekhelet: The Renaissance of a Mitzvah*, Alfred Cohen, ed. (New York: Yeshiva University Press, 1996), pp. 63–78.

12. See Ehud Spanier and Nira Karman, "Muricid Snails and the Ancient Dye Industries," in *The Royal Purple and the Biblical Blue*, Ehud Spanier, ed., *op. cit.*, pp. 179–92.

13. Aristotle, *Historia Animalium*, D'Arcy Wentworth Thompson, trans. (Oxford: Clarendon, 1910).

14. Professor Azul's account is based on Otto Elsner and Ehud Spanier, "The Past, Present, and Future of Tekhelet," in *The Royal Purple and the Biblical Blue*, Ehud Spanier, ed., *op. cit.*, pp. 167–77. See also I. Irving Ziderman, "Purple Dyes Made from Shellfish in Antiquity," *Review of Progress in Coloration* (1986) 16:46-52. Dr. Ziderman traces the rediscovery of the dye back earlier, to Bartolommeo Bizio.

15. There is no agreement on this. According to Baruch Sterman, one of the participants in the recent revival of *tekhelet* production, most rabbis favor the bluer (less purple) color.

16. See Heinrich Zollinger, "Welche Farbe hat der antike Purpur?," *Textilveredlung (1989)* 24(6): 207–12, and Ziderman, *op. cit.*

17. For an interesting comparative chemical analysis of contemporary *Murex trunculus* dye and an archaelogical purple residue on a seventh-century B.C.E. ceramic basin shard, see Zvi C. Koren, "High-Performance Liquid Chromatographic Analysis of an Ancient Tyrian Purple Dyeing Vat from Israel," *Israel Journal of Chemistry* (1995) 35: 117–124. Dyes from the two sources are nearly identical—they are mixtures of indigo, mono- and dibromo-indigo, and a fourth colorant, which may be dibromoindirubin.

18. For a leading reference to what we know of the chemistry and biochemistry of indigo, see Patrick E. McGovern and Rudolph H. Michel, "Royal Purple Dye: The Reconstruction of the Ancient Mediterranean Industry," *Accounts of Chemical Research* (1990) 23: 152–58.

19. Inge Bösken Kanold, an artist in Lacoste, France, has been able to make *tekhelet* blue and Tyrian purple from snails fished up in the most polluted part of the bay of Marseilles.

20. See Anthony S. Travis, *The Rainbow Makers: The Origin of the Synthetic Dyestuffs Industry in Western Europe* (Bethlehem: Lehigh, 1993).

21. It is legitimate to ask about the environmental consequences of large industrial production. Interestingly, it is estimated that to make from natural (plant) indigo the equivalent of the 14,000 tons of indigo now produced annually would take around 1.4 billion pounds of indigo leaves. About 98 percent of the leaves would become waste in the process. Roy Smith and Sue Wagner, "Dyes and the Environment: Is Natural Better?," *American Dyestuff Reporter* (September 1991): 9–34.

22. The book is ably edited by Ehud Spanier, a marine biologist at the Centre for Maritime Studies at the University of Haifa, who also is the co-author of several of the accompanying studies. These range over topics as diverse and relevant as the textiles and dyes of antiquity, the modern biology of muricid snails, and the consequences to religious observance of reviving the *tekhelet*-making industry.

As an example of the fascinating biology that Spanier and co-workers cite for the muricid snails, let me mention the following. The rock murex, a determined carnivore, employs several feeding techniques. The murex have a rasping organ, called a radula, bearing chitinous teeth, and an accessory boring organ (ABO), among a true arsenal of other aggressive chemical weapons. The murex mounts a snail and begins to rasp a hole through its shell by alternate use of the ABO and the radula. The ABO seems to produce a substance that loosens and softens the surface crystals of the calcium carbonate of the shell and the radula then mechanically scrapes the softened shell away. After boring a neat hole (as small as 0.5 mm in diameter) in the other snail's shell, the murex inserts its proboscis and seems to inject a muscle relaxant into its prey, which the murex then consumes at leisure.

23. Quoted from Gösta Sandberg, *op. cit.*, p. 26. The source is *Papyrus Graecus Holmiensis,* O. Lagercrantz, ed. (Uppsala: Akademiska Bokhandeln, 1913).

24. From Cajsa Warg, *Hjälpreda i Hushällningen för unga Fruentimmer* (Stockholm, 1762). See a longer excerpt translated in Gösta Sandberg, *op. cit.*, p. 171.

25. R. M. Smith, J. J. Brophy, G. W. K. Cavill, and N. W. Davies, "Iridodials and Nepetalactone in the Defensive Secretion of the Coconut Stick Insects, *Graeffea crouani,*" *Journal of Chemical Ecology* (1979) 5: 727; Thomas Eisner, "Catnip: Its Raison d'Etre," *Science* (1964) 146: 1318–20. See also the bufalins, from the skin of toads, fireflies, and a small group of plants: T. Eisner, D. F. Wiemer, L. W. Haynes, and J. Meinwald, "Lucibufagins:

Defensive Steroids from the Fireflies *Photinus ignitus* and *P. marginellus* (Coleoptera: Lampyridae)," *Proceedings of the National Academy of Sciences, USA* (1978) 75: 905, and K. Nakanishi, T. Goto, S. Itô, S. Natori, and S. Nozoe, eds. *Natural Products Chemistry*, Vol. 1 (Tokyo: Kodansha, 1974), pp. 469–75.

26. Nahmanides, in *Ramban: The Disputation at Barcelona*, by Charles B. Chavel, trans & annot. (New York: Shilo, 1983), p. 21.

27. As quoted by R. Isaac Herzog in his University of London 1913 D. Litt. thesis, "Hebrew Porphyrology," in *The Royal Purple and the Biblical Blue*, Ehud Spanier, ed., *op. cit.*, p. 116.

28. *Ibid.*, p. 117.

29. See discussion of hair covering and wigs in Chapter 1.

30. The translation is from I. Ziderman "Halakhic Aspects of Reviving the Ritual Tekhelet Dye in the Light of Modern Scientific Discoveries," in E. Spanier, ed., *The Royal Purple and the Biblical Blue, op. cit.*

31. H. J. Buser, D. Schwarzenbach, W. Petter, and A. Ludi, "The Crystal Structure of Prussian Blue, $Fe_4[Fe(CN)_6]_3 \cdot xH_2O$", *Inorganic Chemistry* (1977) 16: 2704–10. The complicated story of the extent of potassium and water incorporation in Prussian blue is related in this paper; there is an "insoluble" Prussian blue without potassium, a "soluble" one with potassium (that's the one Cora draws).

32. A Yiddishism for a necessary but, to some, unmentionable part of human anatomy.

33. See P. Pilleboue and O. Melnyk, "La Découverte du Blue de Prusse," *L'actualité Chimique* (1991): 289–91.

34. The authors' construction of Cora's hypothesis derives from the work of Israel Irving Ziderman, "Blue Thread of the Tzitzit: Was the Ancient Dye a Prussian Blue or Tyranian Purple?," *J. Society Dyers and Colourists* (1981) 97:362–364.

35. Roald Hoffmann, unpublished.

36. Deuteronomy 22:10.

37. Leviticus 19:19.

38. Leviticus 19:9, Deuteronomy 24:19.

39. Deuteronomy 15:19.

40. Deuteronomy 18:4.

41. Numbers 18:14.

42. The *midrash* (Shoher Tov) quoted here comes from Hayyim Nahman Bialik and Yehoshua H. Ravnitzky, *The Book of Legends (Sefer Haggadah)*, William G. Braude, trans. (New York: Schocken, 1992), p. 92.

43. In the Jewish tradition, even a hanged criminal had to be buried properly. The concept of *Kvod haMet* (respect for the dead), which requires immediate and proper burial, is derived from Deuteronomy 21:22–23.

44. For a detailed analysis of the above-cited piece of demagoguery attributed to Korah, see Nehama Leibowitz, *Studies in Bamidbar* (Numbers) (Jerusalem: World Zionist Organization, 1980), pp. 188–191. Leibowitz has a rich and incisive analysis of many aspects of the Korah story in the same volume.

45. Joseph B. Soloveitchik, *Man of Faith in the Modern World. Reflections of the Rav*, Vol. 1, adapted by Abraham R. Besdin (Hoboken: Ktav, 1981), pp. 140–41.

46. Joseph B. Soloveitchik, *op. cit.*, p. 146

47. See Chapter 3.

48. Joseph B. Soloveitchik, *op. cit.*, Vol. 2, p. 30.

49. Joseph B. Soloveitchik, *op. cit.*, Vol. 2, p. 31.

50. For another sensitive account of the symbolism of blue in Jewish history, see Gershom Scholem, "Farben und ihre Symbolik in der jüdischen Überlieferung und Mystik," in *Judaica 3: Studien für jüdischen Mystik* (Frankfurt: Suhrkamp, 1981), pp. 98–150. We are grateful to Inge Bösken Kanold for bringing this article to our attention.

51. Although he was an assimilated Jew, Herzl was familiar enough with the biblical Korah to use him as a symbol of rebellion. The German is *"Gruppe Korah."* See diary entry for August 14, 1897, p. 574.

52. This soliloquy is taken verbatim (except for a change in tense) from the September 6, 1897, entry in *The Complete Diaries of Theodor Herzl*, Raphael Patai, ed., translated by Harry Zohn (New York: Thomas Yoseloff, 1960), p. 588. Herzl used the Hebrew/Yiddish word *broche*. But he also says in the diary that *"Und als ich zum Altar hinaufging . . .* (And when I went up to the altar . . .)" indicating how assimilated he was

53. Perhaps Herzl came around to Wolffsohn's suggestion. Both flags were displayed at the Zionist Congresses. And in his 1902 novel, *Altneuland* [Theodor Herzl, *Old-New Land*, Lotta Levensohn, trans. (New York: Bloch, 1991)], blue and white flags fly over the then only-imagined Jewish state.

54. The United Nations partitioned Palestine in 1947, on November 29, creating the modern Jewish state of Israel.

55. Translation of Poem No. 19 in Yehuda Amichai, *Time*, Ted Hughes and Yehuda Amichai, trans. (New York, Oxford University Press, 1979), p. 21, with modifications by us, authorized by Yehuda Amichai. Our thanks to Oxford University Press for permission to use this poem.

56. Rachel Osher, Geula Abramovitz, Aviram Reuben, and Sasha are entirely imagined, fictional characters.

57. The imagined discussion that follows is informed by the substantive controversy on the status of the formerly Arab territories that came under Israeli rule after 1967, and the potential of their return. That some people call these "occupied" and others call them "liberated" illustrates the deep chasm within Israeli society, a split that is mainly political, but which has religious overtones.

We will please no one. Be that as it may, there are certain realities: deaths, a precarious political settlement between Israel and the Palestinians, and a religious, *halakhically* grounded debate. Signposts to that debate may be found in J. David Bleich, *Contemporary Halakhic Problems* (New York: Ktav, 1983–89), Vol. II, pp. 169–221; Vol. III, pp. 293–305; and Aviezer Ravitsky, *Messianism, Zionism, and Jewish Religious Radicalism*, Michael Swirsky, Jonathan Chipman, trans. (Chicago: University of Chicago Press, 1996).

58. The family of one of the authors (R. H.) was among the founders of Kibbutz Mizra, and active in the Hashomer Hatzair movement in Poland and Israel. The words are Aviram's; the author has only admiration for those socialist settlers and their ideals.

59. For an extensive discussion of the Nahmanides and Maimonides positions, see Bleich, *Contemporary Halakhic Problems, op. cit.,* Vol. II, pp. 169–221.

60. Shalom Carmy and David Singer, "Rav Kook's Contested Legacy," *Tradition* (1996) 30(3): 6–20. Rabbi Kook also had an unusually positive view of science, and an affinity for evolutionary thinking. See the relevant references in the *Encyclopaedia Judaica* under his entry.

61. The blue-and-white Rishon Le-Zion flag of 1885 was the first local one. Some consider the town of Nes Ziona the first to display such a flag as a national symbol; it was unfurled there at an official building dedication in 1891.

62. Talmud, Shabbat 114a.

Chapter 6: Signs and Portents

1. Many exceptions are made to encourage scholarship in Jewish law, *halakhah.* For example, scholars are exempt from certain taxes and responsibilities. See Maimonides, *Mishneh Torah, Laws of Talmud Torah* 6:10 [Hebrew] English translation in Philip Birnbaum, ed., *Maimonides' Mishneh Torah* (New York: Hebrew Publishing Co., 1967), p. 237. For present-day application see A. Lichtenstein, "The Ideology of the Hesder Yeshiva," *Tradition* (1981) 19: 3, note 23.

2. See Christopher Knight, "Composite Views: Themes and Motifs in Hockney's Art," and Anne Hoy, "Hockney's Photocollages," in David Hockney, *David Hockney: A Retrospective* (Los Angeles: Los Angeles County Museum of Art and Harry Abrams, 1988).

3. J. Kirk T. Varnedoe, "In Detail: Gustave Caillebotte's *Streets of Paris*," *Portfolio* (December–January 1979/80) 1(5): 42–46; J. Kirk T. Varnedoe, Marie Berhaut, Peter Galassi, and Hilarie Faberman, *Gustave Caillebotte* (Houston: The Museum of Fine Arts, 1976).

4. Maimonides, *Mishneh Torah, Laws of Homicide and Life Preservation* 8:5 [Hebrew] English translation in Philip Birnbaum, ed., *op. cit.*, p. 237.

5. See Raymond Lee and Alistair Fraser, "The Light at the End of the Rainbow," *New Scientist* (1 September 1990): 40–44, and references therein. For marvelous representations and explanations of rainbows and related natural phenomena, see David K. Lynch and William Livingston, *Color and Light in Nature* (Cambridge: Cambridge University Press, 1995) pp. 103–121.

6. Nahmanides, *Commentary on Genesis*, Vol. 1, Charles Ber Chavel, trans. (New York: Shilo, 1973), pp. 136–38.

7. *Ibid.*

8. *Ibid.*

9. Nehama Leibowitz, *Studies in Bereshit* (*Genesis*) (Jerusalem: World Zionist Organization, 1986), pp. 86–87. See also Gershom Scholem, *Judaica 3, Studien zur jüdischen Mystik* [German] (Frankfurt am Main: Suhrkamp, 1981), pp. 98–151.

10. For readings on this field of mathematics, see Lofti A. Zadeh, *Fuzzy Sets and Applications: Selected Papers*, R. R. Yager, ed. (New York: Wiley, 1987); Waldemar Karkowski and Anil Mital, *Applications of Fuzzy Set Theory in Human Factors* (Amsterdam: Elsevier, 1986).

11. Umberto Eco, "How Culture Conditions the Colors We See," in *On Signs*, Marshall Blonsky, ed. (Baltimore: The Johns Hopkins University Press, 1986), pp. 157–75.

12. Not everyone sees colors; for a wonderful, sensitive account of the world of achromatopes, see Oliver Sacks, *The Island of the Colour-blind* (London: Picador, 1996).

13. On categorizing colors (and other things), see the eminently readable article by Roy R. Behrens, "On Slicing the Cheese and Treating the Menu Like Stew: Creativity and Categorization," *Journal of Creative Behavior* (1976) 10: 227–38.

14. Sasha Weitman, "National Flags," *Semiotica* (1973) 8(4): 328–67.

15. Maimonides, *Mishneh Torah, Laws of Judges*, Ch. 2 [Hebrew]. English translation in Philip Birnbaum, ed., *op. cit.*, p. 298.

16. Roald Hoffmann and Pierre Laszlo, "Representation in Chemistry," *Diogenes* (1990) 147: 23–51.

17. Of the usual kind, in the sky. One could still see rainbow-like features in other manifestations of light and matter, for instance in oil slicks. See Lynch and Livingston, *op. cit.*

18. For a highly instructive discussion of seeing in science see Ian Hacking, *Representing and Intervening* (Cambridge: Cambridge University Press, 1983), especially Chapter 11, "Microscopes."

19. See Simon Schama, "California Dreamer," *The New Yorker* (March 18, 1996), p. 100. Schama says:

> Hockney would therefore devote himself to inventing forms that would seize the most slippery and momentary episodes of contemporary experience and trap them in his icons: the forever-running shower; the fleeting smile; the permanently suspended splash.

20. The drawing on the left is of a water purification system from Felix B. Ahrens, ed., *Sammlung Chemischer und Chemisch-Technischer Vorträge*, Vol. 5 (Stuttgart: Ferdinand Enke, 1900), p. 177. The drawing at right is of a gas analysis system from Adolphe Carnot, *Traité d'Analyse des Substances Minérales*, Vol. 1 (Paris: Vve. Ch. Dunod, 1848), p. 943.

21. Roald Hoffmann, "Modes of Representation," in *Gaps and Verges* (Orlando: University of Central Florida Press, 1990), pp. 55–56.

22. For an illuminating discussion of the interaction of graphic media and art, see William M. Ivins, Jr., *Prints and Visual Communication* (Cambridge: The M.I.T. Press, 1969).

23. Joseph Naveh, *Early History of the Alphabet* (Jerusalem: Magnes Press, 1982), p. 181.

24. For an excellent introduction to chemistry, see Peter W. Atkins, *Molecules* (New York: Scientific American Library, 1987).

25. For a leading reference to simulations of water structure, see David L. Beveridge et al., "Monte Carlo Computer Simulation Studies of the Equilibrium Properties and Structure of Liquid Water," in *Molecular Based Study of Fluids*, J. M. Haile, and G. A. Mansoori, eds., *Advances in Chemistry Series No. 204* (Washington, D.C.: American Chemical Society, 1983), pp. 297–351, and William L. Jorgensen, "Monte Carlo Results for Hydrogen Bond Distributions in Liquid Water," *Chemical Physics Letters* (1980), 70: 326–329.

26. For a perceptive analysis of students' troubles with the semiotics of chemical symbolism, specifically in the context of water, see Michael J. Strauss, "Mistaking the Map for the Territory," *Journal of College Science Teaching* (1996) 25: 408–12.

27. See Roald Hoffmann, Vladimir I. Minkin, and Barry K. Carpenter, "Ockham's Razor in Chemistry," *Bulletin de la Societé Chimique de France* (1996), 133: 117–30.

28. See, for instance, Terence Hawkes, *Structuralism and Semiotics* (Berkeley: University of California Press, 1977), pp. 123–35.

29. The term "icon" was used by the medieval Franco-Jewish Bible commentator, Rashi (1040–1105) in explaining Genesis 37:2 by saying "the iconic appearance of Joseph was similar to Jacob." The Hebrew word Rashi uses is, in fact, *ikonin*, from the Greek, as ours.

30. Robert K. Merton, *On the Shoulders of Giants* (New York: Harcourt, Bracen, and World, 1965).

31. Martin Krampen, *Geschichte der Strassenverkehrszeichen* (Tübingen: Stauffenburg, 1988). This book has many valuable references to other semiotic discussions of traffic signs.

32. Paul Klee, *Pedagogical Sketchbook* (London: Faber and Faber, 1968), p. 54. See also Carolyn Langer, ed., *Paul Klee* (New York: The Museum of Modern Art, 1987) (*Eros, Mask of Fear*) and Jürgen Glaesemer, *Paul Klee, Die Farbigen Werke im Kunstmuseum Bern* (Bern: Verlag Kornfeld, 1976) (*Conqueror*).

Instructive discussions of Klee's work in the context of this article may be found in Mark Rosenthal, "Deciphering Klee," *Portfolio* (1979–80) 1, No. 5: 62–70; Katalin de Walterskirchen, *Paul Klee* (New York: Rizzoli, 1975); Richard Verdi, *Klee and Nature* (New York: Rizzoli, 1985); Félix Thürlemann, *Paul Klee; Analyse Sémiotique de Trois Peintures* (Lausanne: L'Age d'Homme, 1982).

33. On July 19, 1937, the Nazi government opened in Munich an exhibition called *"Entartete Kunst,"* or "Degenerate Art." It was a remarkable assembly of mostly marvelous (and some schlocky) art. Jumbled together on the walls of the Archaelogical Museum were abstract paintings, Expressionist works, social criticism, and art by Jews and Communists— distinguished only by just that, that the work was by Jews and Communists. Paul Klee was well represented in the show, with 17 works (some with nice arrows in them).

See Stephanie Barron, ed., *Degenerate Art: The Fate of the Avant-Garde in Nazi Germany* (Los Angeles: Los Angeles County Museum of Art, Harry N. Abrams, 1991).

34. Samson Raphael Hirsch, *The Collected Writings, Vol. III, Basic Guidelines for a Jewish Symbolism* (New York: Feldheim Publishers, 1984), translated from the original 1856 essay; Samson Raphael Hirsch, *Horeb*, Vol. I (London: Soncino Press, 1962), pp. cvii–cxx.

35. Samson Raphael Hirsch, *Commentary on the Pentateuch*, E. Oratz, ed. (New York: Judaica Press, 1986), pp. 48–49.

36. For an up-to-date definition of this useful word, see Leo C. Rosten, *The Joys of Yiddish* (New York: Pocket Books, 1970).

37. Moses Maimonides, *The Guide for the Perplexed*, translated from Arabic to Hebrew by Yosef Kappah (Jerusalem: Mossad Harav Kook Publishers,

1973), Part III, Ch. 26. English translation by M. Friedlander (New York: Dover Publications, 1956), p. 311.

38. For a discussion of allegorical interpretations, see Nehama Leibowitz, *Studies in Shemot (Numbers)* (Jerusalem: World Zionist Organization, 1986), pp. 497–507.

39. Eileen C.Schwab and Howard C. Nusbaum, eds., *Pattern Recognition by Humans and Machines* (New York: Academic Press, 1986), and references therein. And computers have come to the aid of Torah checkers: "CompuTora," *The Jerusalem Post* (September 22, 1991), p. 12.

40. See also Maimonides, *Mishneh Torah, Laws of Sefer Torah,* Ch. 8 [Hebrew]. English translation in Philip Birnbaum, ed., *Maimonides' Mishneh Torah* (New York: Hebrew Publishing Co, 1967), pp. 56–58.

41. This was the actual verdict and sentence in Criminal Case #676694/90, handed down July 17, 1990, as can be seen in the protocol of the trial.

42. This chapter was originally published in slightly different form in *Diacritics* (1991) 21(1): 2–23. We are grateful to the editors of *Diacritics* for their consideration and assistance.

Chapter 7: Pure/Impure

1. For most of the Bible passages we have used the translation of *The Holy Scriptures* (Philadelphia: Jewish Publication Society, 1978). One exception is our translation of the Hebrew word *nehoshet*, which can mean copper ore, bronze, or brass. We have followed the *Biblical Encyclopedia* in translating it as bronze everywhere except when it is clear from the context that it refers to copper ore in the earth.

2. Herbert C. Hoover and Lou H. Hoover in footnotes and historical notes in their translation of Georgius Agricola, *De Re Metallica* (New York: Dover Publications, 1950). See especially pp. 362, 390–93, 465–66.

3. For some leading references to ancient metallurgy, see: Robert Maddin, ed., *The Beginning of the Use of Metals and Alloys. Papers from the Second International Conference on the Beginning of the Use of Metals and Alloys, Zhengzhou, China, October 1986* (Cambridge: MIT Press, 1988); Leslie Aitchison, *A History of Metals*, Vol. 1 (New York: Interscience Publishers, 1960).

4. See George Rapp, "On the Origins of Copper and Bronze Alloying," pp. 20–27 of Maddin, *op. cit.*; James D. Muhly, "The Beginnings of Metallurgy in the Old World," pp. 2–20 of Maddin, *op. cit.*

5. Hoover and Hoover, *op. cit.*, p. 391.

6. Biblical commentators disagree on the explanation for Jeremiah's disparagement of bronze and iron in this metaphor. For instance, Sh. D. Luzzato (nineteenth century, Trieste) views the base metals bronze and iron as less valuable than gold and silver. On the other hand, Rashi (eleventh century, France) interprets the metaphor thusly: The people are slanderers, strong as bronze and iron, inflicting injury upon their fellow man.

For a recent exploration of the metaphor, see Paula McNutt, *The Forging of Israel: Iron Technology, Symbolism, and Tradition in Ancient Society* (Sheffield: Almond Press, 1990).

7. A former Cornell president wrote an influential book relevant to this theme: Andrew Dickson White, *A History of the Warfare of Science with Theology in Christendom* (New York: D. Appleton, 1896).

8. Colette, *The Pure and the Impure* (New York: Farrar, Straus, and Giroux, 1966), p. 174. (Translated by H. Briffault from *Ces Plaisirs,* Paris: 1932. Retitled in French in 1942 as *Le pur et l'impur*).

9. In case you'd like to find out what D&C Red No. XY is, we recommend Daniel M. Marmion, *Handbook of U.S. Colorants for Foods, Drugs and Cosmetics,* 2nd ed. (New York: John Wiley, 1984).

10. Frank N. Kemmer, ed., *The NALCO Water Handbook,* 2nd ed. (New York: McGraw-Hill, 1988).

11. The meanings of "pure" are not exhausted by this discussion. There is a rich discourse on "pure" painting and on "pure" poetry. For poetry this may be seen in the wonderful essay by Robert Penn Warren, "Pure and Impure Poetry," in *Criticism: Foundations of Modern Literary Judgment,* M. Schorer, J. Miles, and G. McKenzie, eds. (New York: Harcourt, Brace, and World, 1958); for art in Clement Greenberg, "Towards a New Laocoon," *Partisan Review,* July–August 1940, reprinted in John O'Brian, ed., *Clement Greenberg. The Collected Essays and Criticism. Volume I. Perceptions and Judgments 1939–1944* (Chicago: University of Chicago Press, 1986), pp. 32–34. (The Greenberg citation comes from an interesting discussion of Anselm Kiefer's painting by Jack Flam, "The Alchemist," *New York Review of Books,* February 13, 1992: 31–36.)

Not to speak of the senses of the word as used by Mary Daly, *Pure Lust* (Boston: Beacon Press); Irwin Silverman, *Pure Types Are Rare* (Westpoint, Conn: Praeger Scientific); Arthur Silverstein, *Pure Politics and Impure Science: The Swine Flu Affair* (Baltimore: The Johns Hopkins University Press, 1981).

12. Matthew 5:8. The translation is from *The New English Bible* (Oxford: Oxford University Press, Cambridge University Press, 1980).

13. As quoted by Thich Nhat Hanh, "Water Rises Too," *Parabola* (1995) 20(1): 31–33.

14. For references to the theological agony of unitarian religions that nevertheless invoke a trinitarian Godhead, see Edward T. Oakes, *Pattern of Redemption: The Theology of Hans Urs von Balthasar* (New York: Continuum, 1994).

15. Following the biblical injunction not to boil a kid in its mother's milk (Exodus 23:19) the Talmud elaborates a complex set of codes on distinguishing meat and milk dishes, and the timing of ingestion of meat and milk products.

16. For more information on *shaatnez* see the Website http://www.she-mayisrael.co.il/orgs/shatnez

17. Mary Douglas, *Purity and Danger: An Analysis of Concepts of Pollution and Taboo* (New York: Praeger, 1966), pp. 54–72.

18. See the discussion of the Douglas thesis in Robert Alter, "A New Theory of Kashruth," *Commentary* (August 1979): 46–52; and Michael P. Carroll, "One More Time: Leviticus Revisited," in Bernhard Lang, ed., *Anthropological Approaches to the Old Testament* (Philadelphia: Fortress Press, 1985), pp. 117–26; Mary Douglas has been recently reconsidering her previous view of the purity rules of Leviticus: Mary Douglas, "Figleaves and Other Coverings; Reconsidering the Purity Rules of Leviticus," unpublished; Mary Douglas, "A Mouse, a Bird and Some Fish," to be published; lectures in Jerusalem, November 1996.

19. See also Walter Houston, "Purity and Monotheism: Clean and Unclean Animals in Biblical Law, " *Journal for the Study of the Old Testament, Supplement Series 140* (Sheffield, England: Sheffield Academic Press, 1993).

20. Douglas, *Purity and Danger*, p. 99.

21. What is particularly intriguing is that the notion about separations voiced in the blessing is intimately related to, and in fact begins with, a statement about intelligence. In explaining this, the Jerusalem Talmud asks rhetorically, "If there is not wisdom, how can there be differentiation?" This passage is inserted into the prayer for intelligence, *binah*, in the Saturday night service. Nosson Scherman, ed. and trans., *The Complete ArtScroll Siddur* (Brooklyn: Mesorah Publishing, 1984).

22. For an interesting discussion of another passage from the Prophets concerned with ritual purity, see Victor Hurowitz, "Isaiah's Impure Lips and Their Purification in Light of Akkadian Sources," *Hebrew Union College Annual* (1989) 60: 39–89.

23. A variety of views on the rationale for the forbidden foods are contrasted by Nehama Leibowitz, "The Dietary Laws," in *Studies in Vayikra (Leviticus)*, A. Newman, trans. (Jerusalem: World Zionist Organization, 1983), pp. 76–86.

24. It cannot be emphasized enough that the concept of *taharah* (purity) is one of the most complex and baffling aspects of Jewish law. The section of the Mishnah entitled "Purities" (containing several tractates) is the longest of the six orders, and in his commentary on it, Maimonides warns that:

"I wrote this introduction to disabuse you of the notion that these laws are like those of the Feast of Tabernacles, or Judges' Oaths, so that you will not read the laws of Purities and think that you have grasped them on first sight. . . . These are among the most abstruse of the Talmud, being difficult even for the greatest sages. The only way to master them is to expend days and lose nights, and to progress as one accumulates penny by penny a great fortune."

25. Wendy Doniger, "Why Should a Priest Tell You Whom to Marry? A Deconstruction of the Laws of Manu," *Bulletin of the AAAS* (March 1991) 44(6): 18–31.

26. Midrash Tanhuma, Parashat Shemini.

27. George Frideric Handel, *The Messiah,* Section No. 7–Chorus, "And He Shall Purify." Other metal metaphors are used in the oratorio: Section No. 6 is also from Malachi 3, "For he is like a refiner's fire," and No. 43, "Air for Tenor," based on Psalm 2:9, "Thou shalt break them with a rod of iron." The illustration shown is in the hand of John Christopher Smith, Handel's principal copyist. We are grateful to Peter A. Ward Jones of the Bodleian Library for his instructive comments.

28. Nicolás Guillén, "I Declare Myself an Impure Man" [*Digo que yo no soy un hombre puro*], in *¡Patria y Muerte!* [*The Great Zoo and Other Poems*], R. Márquez, ed. and trans. (Havana: Editorial de Arte y Literatura, 1972), pp. 210–13.

29. For an introduction to the marvels of photosynthesis, see Lubert Stryer, *Biochemistry,* 4th ed. (New York: W. H. Freeman, 1995), Ch. 26.

30. Günther Ohloff, *Perfumer and Flavorist* (1978) 3: 11.

31. That there is a relentless natural tendency to disorder has been disputed, most eloquently by Ilya Prigogine. In fact, the English title of his book with Isabelle Stengers is *Order Out of Chaos*. Prigogine agrees with the classical view that in closed systems (no transfer in or out of matter or energy) approaching equilibrium, entropy maximization reigns. But he and Stengers argue that in open systems, far from equilibrium, "new types of structures may originate spontaneously. In far-from-equilibrium conditions we may have transformations from disorder, from thermal chaos, into order." [Ilya Prigogine and Isabelle Stengers, *Order Out of Chaos: Man's New Dialogue with Nature* (New York: Bantam, 1988), p. 12]. That is so, but only locally, over a finite period of time, and not without a disordering cost elsewhere in the universe. Prigogine's ideas are stimulating and thought-provoking. But they are easily overstretched, to give us what our minds desire, that order emerging out of chaos is natural. [See also the review of the Prigogine and Stengers book by Heinz R. Pagels, "Is the Irreversibility We See a Fundamental Property of Nature?" *Physics Today* (January 1985): 97–9, and Rolf Landauer, "Nonlinearity, Multistability and Fluctuations: Reviewing the Reviewers," *American Journal of Physiology* (1981) 241: R107–R113.]

32. A most readable account of thermodynamics is to be found in Peter W. Atkins, *The Second Law* (New York: Scientific American Books, 1984).

33. For some related musings, see Roald Hoffmann, Vladimir I. Minkin, and Barry K. Carpenter, "Ockham's Razor in Chemistry," *Bulletin de la Societé Chimique de France* (1996) 133: 117–30.

34. Thomas Eisner, Letter [to Roald Hoffmann], 18 March 1991.

35. Of course it takes six for sex: L. B. Bjostad, C. E. Linn, J.-W. Du, and W. L. Roelofs, "Identification of New Sex Pheromone Components in *Trichoplusia ni*, Predicted from Biosynthetic Precursors," *Journal of Chemical Ecology*, (1984) 10(9): 1309–23. Wendell Roelofs and Thomas Glover, "Genetics of a Moth Pheromone System," Ch. 9 in *Chemical Senses*, Vol. 3, C. J. Wysocki and M. R. Kare, eds. (New York: Marcel Dekker, 1991). Charles E. Linn and Wendell Roelofs, "Pheromone Communication in the Moths and Its Role in the Speciation Process," in D. M. Lambert and H. G. Spencer, eds., *Speciation and the Recognition Concept* (Baltimore: The Johns Hopkins University Press, 1995), pp. 263–300.

36. Robert M. Silverstein, "Pheromones: Background and Potential for Use in Insect Pest Control," *Science*, 213: 1326–32, and references therein.

37. Robert L. Stevenson, *Dr. Jekyll & Mr. Hyde, the Merry Men and Other Tales* (London: J. M. Dent, 1925), p. 61. This quotation was brought to our attention by a beautiful column by David Jones, which makes some of the same points this chapter does: David Jones, "Impure Thoughts," *Chemistry in Britain* (1992) 28: 928.

38. See notes 3 and 4.

39. The data in the figure is from *Copper Data 1936* (London: British Copper Development Assoc., 1937).

40. For a leading reference, see Ian Baker, "Metals Pass the Endurance Test," *New Scientist* (May 30, 1992): 34–38; also J. E. Gordon, *The Science of Structures and Materials* (New York: Scientific American Library, 1988), Ch. 5.

41. Samuel Noah Kramer, *The Sumerians: Their History, Culture, and Character* (Chicago: University of Chicago Press,1963), pp. 264–65.

42. The original tablet is in the University Museum, University of Pennsylvania.

43. We are grateful to Professor Yaakov Klein, of Bar-Ilan University, for introducing the authors to the world of Sumer and for his thorough investigation of whether the copper in the dispute is pure copper or an alloy.

Professor Klein personally is of the opinion that the copper in the disputation presented represents the basic, pure, copper, from which all objects, whether of pure copper or of its alloys, were manufactured. Similarly, its opponent, the silver, seems to represent all objects made from pure silver or its alloys.

Note, however, that the term used for copper throughout this dialogue is *urudu-kal-ga*, copper the strong one, which may be a synonym for *zabar*, bronze. See Henri Limet, *Le Travail du Métal au Pays de Sumer, Bibliotheque de la Faculté de Philosophie et Lettres de l'Université de Liège*, Vol. 155 (Paris: Les Belles Lettres, 1960).

44. See note 41.

45. Cyril Stanley Smith, "The Discovery of Carbon in Steel," Ch. 2 in *A Search for Structure: Selected Essays on Science, Art, and History* (Cambridge: MIT Press, 1981). First published in *Technology and Culture* (1964) 5: 149–75 .

46. Smith, *op. cit.*, p. 35.

47. See William H. Brock, *The Norton History of Chemistry* (New York: W. W. Norton, 1992).

48. John Pettus, *Volatiles from the History of Adam and Eve* (London: T. Bassett, 1674). We are grateful to Cyril Smith for bringing this remarkable text to our attention.

49. Mircea Eliade, *The Forge and the Crucible*, S. Corrin, ed. (Chicago: University of Chicago Press, 1962), p. 56.

50. See Roald Hoffmann, *The Same and Not the Same* (New York: Columbia University Press, 1995), especially Ch. 47– Ch. 50.

51. Robert Hazen, "Perovskites," *Scientific American* (June 1988): 52–61. Robert J. Cava, "Superconductors beyond 1-2-3," *Scientific American* (August l990) 263(2): 42–49; R. Simon and A. Smith, *Superconductors Conquering Technology's New Frontier* (New York: Plenum, 1988).

52. Thomas Hellman Morton, Letter [to Roald Hoffmann], 12 December 1990.

53. Some excellent examples of the necessity of precisely defined mixtures are to be found in perfumery: Édouard Demole, "Parfums et Chimie: Une Symbiose Exemplaire," *L'Actualité Chimique* (May–June 1992): 227–37.

54. Aldous Huxley, *Music at Night* (London: Chatto & Windus, 1931). This quotation, aptly applied to Gustav Mahler's music by Deryck Cooke [in Deryck Cooke, *Gustav Mahler, An Introduction to his Music* (Cambridge: Cambridge University Press, 1980)] was kindly brought to our attention by Brenda Miller.

55. This paragraph from the Jewish daily morning service originates in the Talmud, Kritot 6a, where the metaphor involving galbanum also appears.

56. Purity is not a simple concept. While from a physical or chemical viewpoint the incense is patently a mixture, from the ritual perspective it is "pure and holy" (Exodus 30:35).

57. According to A. Varvoglis, University of Thessaloniki, a Holy Myrrh of fifty-seven ingredients is prepared ceremonially every decade by the Patriarchate of Constantinople, and then distributed world-wide. A thirty-eight

component version is specified in the *Mega Euchologian*, the Greek Orthodox expanded prayer book.

58. See note 55.

59. Maimonides, *Introduction: Chapter 6, Commentary on the Mishnah*
[Hebrew] (Jerusalem: Mosad Harav Kook, 1978).

60. For an introduction to this subject, see A. Steinsaltz, *The Essential
Talmud* (New York: Basic Books, 1987), Ch. 22.

61. The Delaney Clause occasioned impassioned debate since its 1958
passage. For a good introduction to its history and currency, see the testimony in: United States Congress, House Committee on Government Oper-
ations, Human Resources and Intergovernmental Relations Subcommit-
tee, *FDA's Regulation of Carcinogenic Additives* (Washington, D.C.: U.S.
Government Printing Office, 1977).

  In practice, the clause was effectively replaced by a "negligible risk
standard"—defined as leading to not more than one additional case of
cancer per million people. In August 1996 a pesticide reform bill was
passed, which abandoned the absolutism of the Delaney Clause. The consequences of the new standards are not yet clear: David Hanson, "Regulation After Delaney," *Chemical and Engineering News* (September 23, 1996),
pp. 38–39.

62. Tobias Geffen, "A *Teshuvah* Concerning Coca-Cola," Louis Geffen and M.
David Geffen, trans., Joel Ziff, ed, in *Lev Tuviah: The Life and Works of Rabbi
Tobias Geffen* (Newton, Mass.: Geffen Memorial Fund, 1988), pp. 117–21.

63. Most interesting in the talmudic argument is the development of a
calculus of probabilities and statistical inference. A classic case is the
problem of a liquid mixture of a forbidden substance with a permitted one
(e.g., milk in a meat broth; forbidden animal oil into permissible oil). If a
sample is taken, can it be inferred that when actual mixing occurs, "there
is homogeneity" (Hebrew: *yesh bilah*) throughout and therefore the proportions of the sample reflect exactly the proportions of the original? How
large does the sample have to be with reference to homogeneity and are
the rules different for liquids and solids? To learn the rabbis' solution to
the problematic solution you would have to consult the Talmud, or
Nachum L. Rabinovitch, "Variability in Samples," in *Probability and Statistical Inference in Ancient and Medieval Jewish Literature* (Toronto: University of Toronto Press, 1973), p. 82.

64. Douglas, *Purity and Danger, op. cit.*, p. 162.

65. As might have been expected, our legal systems also struggle with the
notion of purity. Does the purification of a known, patented mixture result
in a product that is newly patentable? One line of legal reasoning holds
that it does, another focuses on whether there is a difference "in kind"
rather than "in degree" in the utility of the newer compound. See Donald
S. Chisum, *Patents*, Vol. 1 (New York: Matthew Bender, 1978), pp. 53–57,

1.02 [9]. Also *Amgen Inc. v. Chugai Pharmaceutical Co. Ltd.*, 13 U.S.P.Q. 2d 1737, 1759 (D. Mass. 1989). We are grateful to C. Frederick Leydig for bringing these cases to our attention.

66. Tacitus, *Germania,* 4–6, H. Mattingly, trans. (Hammondsworth: Penguin, 1970), 104–05.

67. Michel Tournier, *The Ogre,* Barbara Bray, trans. (Garden City: Doubleday, 1972), p. 75. The French original is entitled *Le Roi des Aulnes* (Paris: Gallimard, 1970), p. 85.

68. Primo Levi, *The Periodic Table,* R. Rosenthal, trans. (New York: Schocken, 1984), pp. 33–4.

69. This chapter was published in slightly different form in *The New England Review* (1994) 16(1): 41–64. We are grateful to the editors of *The New England Review,* © Middlebury College Publications, for their editorial assistance.

Chapter 8: Camel Caravans in the Pentagon

1. Ann Tusa and John Tusa, *The Nuremberg Trial* (New York: Atheneum, 1984) p. 321; H. J. Zimmels, *The Echo of the Nazi Holocaust in Rabbinic Literature* (New York, Ktav, 1977), Ch. 4

Max Weinreich, *Hitler's Professors: The Part of Scholarship in Germany's Crimes Against the Jewish People* (New York: Yiddish Scientific Institute—YIVO, 1946), pp. 97–113.

2. Early on in the Nazi regime, "harmful" books were confiscated and publicly burned. See *Das war ein Vorspiel nur . . . Bücherverbrennung Deutschland 1933* (Berlin: Medusa, 1983).

3. Solomon B. Freehof, *On the Collecting of Jewish Books* (New York: Society of Jewish Bibliophiles, Munro Press, 1961). In this section we draw heavily on Rabbi Freehof's inspired account.

4. This anonymous composition is discussed by Ismar Schorsch in *Moritz Oppenheim, The First Jewish Painter* (Jerusalem: Israel Museum, 1983), p. 54.

5. One of the authors (S. L. S.) heard Rabbi Selig Auerbach read on the holiday of Purim from a parchment scroll of Esther he got from his uncle, Michael, a Jew from Mainz, who was in the German army on the Western front in World War I. After Michael's unit overran a French position, he found an unusual Purim scroll (*megillah*) hastily abandoned, perhaps by a French chaplain. The young soldier wrote to his family that he would bring it home on his next leave. Unfortunately, he fell in battle, and his belongings, including the scroll, were shipped to his parents. Selig Auerbach was nearing *bar mitzvah* age, so the *megillah* was given to him. Rabbi Auerbach, now living in Rochester, N.Y., still reads in public from this battlefield scroll on Purim.

6. This hearkens back to Chapter 4, "Bitter Waters Run Sweet," in which we discuss the period when the Israelites trudged three days in the desert without water. One metaphoric explanation hinges on the equation Torah = water. So from that time on the Israelites should never go three days without Torah; therefore the Torah is read Mondays, Thursdays, and Sabbaths. A more prosaic reason is that Ezra and Nehemiah instituted the Monday and Thursday public Torah readings because those were (and still are) the Middle Eastern market days.

7. The regulations regarding the handling of a Torah scroll are stricter than those for a printed Bible. If a Bible is dropped, one picks it up and kisses it; if a scroll is dropped, fasting (for forty days) is called for, but one may give charity in lieu of that. A Bible is opened and read with reverence; a scroll is not touched but is read while using a silver pointer.

8. Zvi Hirsch Hayyot (Chajes), *The Student's Guide through the Talmud*, Jacob Schachter, trans. (London: East and West Library, 1952).

9. Zvi Hirsch Hayyot (Chajes), *Kol Sifrei Maharatz Chajes* [Hebrew] (Tel Aviv: Divre; Hakhamim, 1958), #11.

10. Menahem Azaria de Fano (Rama), *Bar Ilan Responsa Project* (Spring Valley, N.Y.: Torah Educational Software, 1994), Responsa #38, #93, #95; Nisson E. Shulman, *Authority and Community: Polish Jewry in the Sixteenth Century* (New York: Ktav, 1986), Responsum #99; Israel M. Tashma, "Reading 'Our' Pentateuchs—the History of Two Halakhic Approaches" [Hebrew], Ch. 6 in *Early Franco-German Ritual and Custom* [Hebrew] (Jerusalem: Magnes, 1992), Ch. 6, pp. 171–81.

11. The section of the Dead Sea Scrolls is from the Temple Scroll, and is in The Shrine of the Book, Israel Museum, Jerusalem. The letter by Judah Halevi from the Cairo Genizah is now in the Jewish Theological Seminary, New York.

12. Joseph Messas, *Mayyim Hayyim* [in Hebrew] (Fez: Masud Sharvit, 1933), pp. 38–39. The copy we used is from the Freehof collection at the Hebrew Union College.

13. Bartolommeo di Giovanni was active in Florence between 1488 and 1501, working initially with Domenico Ghirlandaio. This painting of Joseph being sold to merchants in a caravan is so vivid you can almost smell the contents of the camels' loads (from their shape we can't be sure if Bartolommeo actually observed camels. . . .) In Genesis 37:25 the brothers, "looking up, saw a caravan of Ishmaelites coming from Gilead, their camels bearing gum, balm, and ladanum to be taken to Egypt." The mention of these delightful aromas seems superfluous. One *midrash* gives us a whiff of what is to come in our tale of leather and parchment. "Now surely Ishmaelites generally carry hides and *itran* (a lighting resin)." Both emit repugnant smells. "But see how the Holy One . . . prepared sacks filled with spices for [Joseph] that the wind might waft their fragrance to him to counteract the foul odors." [See *Midrash Rabbah. Genesis*, Vol. 2, H. Freedman, trans. (London: Soncino Press, 1983), p. 782.]

14. Commission on Jewish Chaplaincy, *Responsa to Chaplains 1948–53* (New York: Jewish Welfare Board, 1953), Responsum #8. The proper Hebrew plural of *Sefer Torah* is *Sifrei Torah*.

15. Our thanks to Bernard Rabenstein, the reference librarian at Hebrew Union College, where the Freehof book collection is housed. Mr. Rabenstein helped us locate the Chaplaincy Committee and the Rabbi Messas responsum.

16. Leila Avrin, *Scribes, Script, and Books* (Chicago: American Library Association, 1991), Ch. 5, Ch. 9. We follow the definitions given by Avrin for vellum, parchment, and leather. However, definitions of these traditional materials are not made in heaven, and so others define them differently. See also the articles on Paper and Leather in Volumes 14 and 16 of the *Kirk-Othmer Encyclopedia of Chemical Technology*, 3rd ed. (New York: Wiley, 1978).

17. Ronald Reed, *The Nature and Making of Parchment* (Leeds: Elmete Press, 1975), p. 27

18. Mishnah, Baba Batra 2:9; see also Talmud, Ketubot 77a, where a woman can press for divorce if her husband has become repulsive to her because he started to work in a tannery.

19. The passage from the *Zohar* of Moses of Léon is from *Zohar: The Book of Enlightenment*, Daniel Chanan Matt, trans. (New York: Paulist Press, 1983), pp. 161–62; see also important notes on pp. 276–77.

20. The originality of the *Zohar* is in the interpretation (see Matt, preceding endnote): The actual verse in Leviticus 26:44 is: "... I will not reject them or spurn them so as to destroy them." The Hebrew word for "to destroy them" is *"le-khallotam."* But the written text of the Torah omits one letter, making the word literally *"le-khalltam."* This allows the *Zohar* a play on words, one with significance, relating the changed word to the phrase "to their bride" (*le-khallatam*).

21. Talmud, Kiddushin 82a, b.

22. The illustration of collagen is by Irving Geis, from Donald Voet and Judith G. Voet, *Biochemistry* (New York: Wiley, 1990); further details of the structure and mode of action of this incredible biopolymer may be found on pp. 159–64.

23. To use the textile chemists' notation (see Chapter 3), the individual helices are S twist, and the triple helix is a Z ply or braid.

24. Primo Levi, *The Periodic Table*, Raymond Rosenthal, trans. (New York: Schocken, 1984), pp. 200–10.

25. Benjamin Vorst, "Parchment Making—Ancient and Modern," *Fine Print* (1986) 12(4): 209–211, 220.

26. This is the only case where God's name is permitted to be erased, for the sake of restoring peace and confidence between husband and wife.

27. Avrin, *op. cit.*, p. 115.

28. Chaim Twersky, "The Use of Modern Inks for Sifrei Torah," *Journal of Halacha and Contemporary Society* (1988) 15: 68–76. Israel M. Tashma, "On Ink and the Technique of Parchment Making" [Hebrew], Ch.17 in *Ritual Custom and Reality in Franco-Germany 1000–1350* [Hebrew] (Jerusalem: Magnes, 1996), pp. 289–302.

29. James Burke, *Connections* (Boston: Little, Brown, 1978), pp. 100–101.

30. We paraphrase Abraham Joshua Heschel, who asked a student "Which molecule has the dignity?"; see Joshua Halberstam, "Which Molecule Has the Dignity," *Tikkun* (1994) 9(3): 24.

31. The imperative to beautify the scrolls is expressed in an unexpected context in the discussion in the Talmud, Shabbat 133b allowing the Sabbath to be violated in order to perform a circumcision in a more aesthetic way. The prooftext that is the basis for this leniency is the *midrash* on Exodus 15:2—"'This is my God I will beautify him,' that is, make a beautiful *sukkah* in His honor, a beautiful lulav, a beautiful shofar, beautiful fringes, and a beautiful Torah scroll, and write it with fine ink, a fine reed pen, and a skilled penman, and wrap it about with beautiful silks." Note more than half of the list deals with adorning a Torah scroll. See also Chapter 2, note 17, p. 312.

32. *Kavannah* in Hebrew. Rabbi J. B. Soloveitchik has written that no concrete object possesses sacred status; an object can be sanctified only through man's actions. Judaism abhors fetishism, animism, and any kind of magical approach to the physical world; holiness is obtained only via human actions. Thus the holiness of a Torah scroll does not come automatically from the text, but is infused into the writing by the human efforts of a scribe through his deliberate intention: Joseph B. Soloveitchik, "The Jew Is Like a *Sefer Torah*," trans. from Yiddish to Hebrew by Shalom Carmy, *Bet Yosef Shaul Journal* (1994) 4: 66–100.

33. To be specific, the scribe says: "I declare that I am writing this for the sake of the sanctity of God's name." If the scribe forgets to say the declaration on even one instance of God's name, then the whole scroll is not ritually fit. Since you can't erase God's name (but note the exception we cite of the *sotah*) or scrape it off with a razor in order to rewrite it, how does a scribe deal with this mistake? It's a hard life—if the declaration on God's name were forgotten in one instance, or if a mistake were made in writing it, the only thing to do is deposit that parchment sheet in a *genizah* and write another whole sheet. This is perhaps one reason the writing is done on several dozen small individual skins: If there is a mistake in one, you only have to redo that piece of parchment.

34. Luke 5:37–39. The translation here is from *The Holy Bible: New International Version* (Grand Rapids: Zondervan, 1973).

35. This metaphor appears also in Matthew 9:17 and Mark 2:22. R. H. is grateful to Jonathan Bishop and Sheila D'Atri for a discussion of the Luke line.

36. Quoted, with permission, from a conversation between Jonathan Bishop and Roald Hoffmann.

37. Gladys A. Reichard, *Navajo Religion* (Princeton: Princeton University Press, 1970).

38. This translation of *"Im yesh et nafshekha ladaat"* (Tel Aviv: Dvir Publishers, 1966) is by the authors, who relied on an earlier translation by Harry H. Fein in Israel Efros, ed., *Selected Poetry of Hayyim Nahman Bialik* (New York: Bloch, 1965), pp. 70–72.

39. Jacob Neusner, "Letter to the Student," in *The Classics of Judaism* (Louisville: Westminster John Knox Press, 1995), p. xiv.

40. Neusner, *ibid.*, pp. xiii–xvii.

41. Mishnah, Peah 1:1; Talmud, Shabbat 127a.

42. Ephraim Oshry, *Responsa from the Holocaust*, Y. Leiman, trans. (New York: Judaica Press, 1983), pp. 173–74. By financing the writing of a new scroll the earlier loss would be somewhat ameliorated.

43. See the entry "People of the Book" in Mustansir Mir, *Dictionary of Qur'anic Terms and Concepts* (Garland Publishing: New York, 1987), pp. 153–54.

Epilogue

1. Much has been written about the relationship of science and religion. In the context of the kinship of the logic of Jewish religious thought and science, we find much in common with an article brought to our attention by Cyril Domb: Shmuel Safran's "Methodologies Common to Science and *Halakhah*," *Bekhol Derakhekha Daehu—Journal of Torah and Scholarship* (Bar Ilan University Press, Summer 1996) 3:5–19.

2. Richard Powers, *The Gold Bug Variations* (New York: William Morrow, 1991) p. 611.

3. Yehiel Mikhal of Zloczow, as translated by Martin Buber, *Tales of the Hasidim: The Early Masters*, Olga Marx, trans., (New York: Schocken, 1975), p. 147.

Credits

We gratefully acknowledge the individuals, institutions, and copyright holders listed below for their permission to reproduce images or text in this book.

Front Cover Image: Rolnik Publishers, Something Different, P.O. Box 17075, Tel Aviv, Israel. **Chapter 1, 6: (t, b):** Orville Redenbacher's is a registered trademark of Hunt-Wesson, Inc; used with permission. Reprinted with permission from King Oscar USA; **14(t, b):** Jane Singer; **15:** Detail of wooden panel on door of Basilica Santa Sabina, Rome, fifth century C.E.; reproduced with permission of Curia Generalitia Dominicani Roma; **20:** Oren Scherman, Roald Hoffmann; **21:** Oren Scherman, Roald Hoffmann; **24–25:** poem, "Prayer" by Joseph Bruchac; reprinted with permission of the author; **27:** poem by Judah Halevi, translated by Raymond Scheindlin. Used by permission of the Jewish Publication Society; **27–29:** poems by Solomon ibn Gabirol and Moses ibn Ezra, translated by Raymond Scheindlin; **34–35:** excerpts of lyrics from *Hair*, EMI Music Publishing; **36–40:** text previously published in *Discover* magazine; **37:** Ian Worpole, August *Discover* © 1993; reprinted with permission of Discover Magazine; **39:** Jane Jorgensen, Roald Hoffmann; **40–41, 53–54:** © Barak Epstein, quoted by permission. **Chapter 2, 56:** Oscar and Regina Gruss Collection; **59:** ASAP/Kenneth L. Fischer; **63:** Eli Silberstein, Gary S. Michalec; **65:** Sima H. Morell, ASAP/Kenneth L. Fischer, Gary S. Michalec; **67:** Isaac Yallouz; **69:** Shira Leibowitz Schmidt; **71:** Eli Silberstein. **Chapter 3, 81:** Oren A. Scherman, Norman Goldberg, Roald Hoffmann; **82(t):** Oren A. Scherman, Norman Goldberg, Roald Hoffmann; **82(b):** pages 290 and 294 of *Textile Science*, by Kathryn L. Hatch © 1993 by West Publishing Company. All rights reserved; **83(l):** Oren A. Scherman; **83(r):** detail of God's hand from *The Creation of Adam*, Michelangelo, Sistine Chapel, Vatican Museums; **83:** Roald Hoffmann; **84:** Ronald N. Bracewell, *The Galactic Club: Intelligent Life in Outer Space* (New York: Norton, 1976); **87:** John E. McMurry and Robert C. Fay, *Chemistry* (Englewood Cliffs, NJ: Prentice Hall, 1995); **88:** Klaus Roth, Simone Hoeft-Schleeh, VCH Verlagsgesellschaft. The Pasteur crystals are in the Deutsches Museum, Munich, Germany; **95:** ASAP/Kenneth L. Fischer; **97:** Allen Memorial Art Museum, Oberlin college, Ohio; Mrs. F. F. Prentiss Fund, 1963; **101:** Museum Boerhaave, Leiden, Netherlands; Verlag Helvetica Chimica Acta; Edgar Heilbronner, Jack D. Dunitz; **104(l):** Kunstmuseum Bern, Switzerland, Stiftung Karl und Jürg Im Obersteg; **104(r):** Sprengel Museum, Hannover, Germany. © 1997 Artists Rights Society (ARS), New York/ADAGP, Paris; **106:** lyrics © Shapiro, Bernstein and Co., 1958; **108:** Beatriz Da Costa, New York; **114:** quote © Sam Orbaum; **114:** I. Hargittai and M. Hargittai, *Symmetry: A Unifying Concept*, second printing, Random House, New York, 1996; **121:** Jaap J. Vermeulen. **Chapter 4, 140:** Gary S. Michalec, Shira

Liebowitz Schmidt; **145:** Emmanuel Mazor; **148:** James Case, *Sensory Mechanisms* (New York: Macmillan, 1966); **149:** Norman Goldberg; **153:** "Ordeal of Bitter Waters," by C. N. Schurtz. Reproduced after Rachel Wischnitzer; **Chapter 5, 160:** Israel Defense Forces (I.D.F.) Spokesperson; **168:** Ehud Spanier, Center for Maritime Studies, University of Haifa; **169(t):** Kadman Numismatic Pavilion of the Eretz Israel Museum, Tel Aviv, Israel; **169(b):** Jane Jorgensen, Roald Hoffmann; **171:** Ehud Spanier, Ed., *The Royal Purple and the Biblical Blue* (Jerusalem: Keter, 1987). Courtesy of Yitzhak Herzog; **179:** Shenkar College of Textile Technology and Fashion, Ramat Gan, Israel; **183:** Grigori Vajenine; **193:** Central Zionist Archives, Jerusalem, Israel; **196:** Designed by Sharon Murro for Bezeq, The Israel Telecommunications Corp. Ltd.; **200:** Central Zionist Archives, Jerusalem, Israel; **202–203:** poem by Yehuda Amichai, Oxford University Press, 1979; **210:** photograph by Avi Ohayon, Israel Government Press Office. **Chapter 6, 213–238:** Shira Leibowitz and Roald Hoffmann, *Diacritics*, Volume 21:1, 1991, pp 2–23, © The Johns Hopkins University Press; **214:** Jane Jorgensen, Shira Leibowitz Schmidt; **215:** Jane Jorgensen, Roald Hoffmann; **216:** David Hockney, "Pearblossom Hwy., 11-18th April 1986," #2, photographic collage, 71 × 107", © David Hockney; **217:** The Art Institute of Chicago; **224:** ASAP/Kenneth L. Fischer; **225:** David Hockney, "The Splash," acrylic on canvas, 71 × 72", © David Hockney; **226(l):** Felix B. Ahrens; **226(r):** Adolphe Carnot; **228(t):** Norman Goldberg, Roald Hoffmann; **228(b):** Donald B. Boyd, Eli Lilly and Co.; **229:** S. Swaminathan and D. L. Beveridge, *J. Am. Chem Soc.*, **99:**7817 (1978); **233:** Jane Jorgensen, Roald Hoffmann; **237:** Central Zionist Archives, Jerusalem, Israel. **Chapter 7, 239–262:** previously published in modified form in *The New England Review* © Middlebury College Publications; **242:** Warshaw Collection of Business Americana, Archives Center, National Museum of American History, Smithsonian Institution; **245:** Courtesy of the Bodleian Library, Oxford, United Kingdom; **249:** *Physical Metallurgy for Engineers 2/e*, by Clark and Varney, © 1962 by Van Nostrand Reinhold; reprinted with permission from Wadsworth publishing Co.; **251:** University Museum, University of Pennsylvania, Philadelphia; **253:** Daniel Stoltzius von Stoltzenberg, *Chymisches Lustgärtlein*. Reprinted in 1968 by Wissenschaftliche Buchgesellschaft, Darmstadt, Germany; **254:** Jane Jorgensen, Roald Hoffmann; **259:** Joel Ziff, Rabbi Tobias Geffen Memorial Fund. **Chapter 8, 264:** Gary S. Michalec, Shira Leibowitz Schmidt; **265:** Bildarchiv Preussischer Kulturbesitz, Berlin, Germany; **267:** The Israel Museum, Jerusalem, Israel; **270(t):** The Israel Museum, Jerusalem, Israel; **270(b):** The Jewish Theological Seminary of America; **272:** Fitzwilliam Museum, Cambridge, United Kingdom; **274:** After *The Nature and Making of Parchment* by Ronald Reed, © 1975 by the Elmete Press; reproduced by permission; **276:** Copyright by Irving Geis; **277:** Bernard Gotfryd; **280:** Leila Avrin; **285:** Baruch Schmidt; **286–287:** excerpt from poem by Hayyim Nahman Bialik; Dvir Publishers, Tel Aviv, Israel; **289:** ASAP/Kenneth L. Fischer.

Color Plates

Plate 1: Tnuva Dairies; Gitam Advertising, Israel. **Plate 2:** Director and University Librarian, The John Rylands University Library of Manchester, United Kingdom. **Plate 3:** Stichting Vrienden van het Mauritshuis, The Hague, Netherlands. **Plate 4:** National Palace Museum, Taipei, Taiwan. **Plate 5:** Yaffa Wigs, Inc., Brooklyn, N.Y., and Bnei Brak, Israel. **Plate 6:** Shira Leibowitz Schmidt. **Plate 7:** Solgar Vitamins, Symbol-Peres Advertising & Marketing, Israel. **Plate 8:** ASAP/Kenneth L. Fischer. **Plate 9:** Ehud Spanier, Center for Maritime Studies, University of Haifa; D. Darom. **Plate 10:** Shira Leibowitz Schmidt. **Plate 11:** Sharon Kopelnikov and Efrat Leibowitz; Museum of the History of Rishon Le-Zion, Historical Archives, Rishon Le-Zion, Israel. **Plate 12:** Beth Hatefutsoth Museum, Tel Aviv, Israel. From the collection of Hayim Shtayer, Haifa. **Plate 13:** Shira Leibowitz Schmidt. **Plate 14(l):** David K. Lynch and William Livingston; **Plate 14(b):** United States Postal Service. **Plate 15:** Paul Klee Foundation, Kunstmuseum Bern, Switzerland. © 1997 Artists Right Society (ARS), New York/VG Bild-Kunst, Bonn, Germany. **Plate 16:** The Museum of Modern Art, New York. © 1997 Artists Right Society (ARS), New York/VG Bild-Kunst, Bonn, Germany. **Plate 17:** Collection Angela Rosengart, Lucerne. © 1997 Artists Right Society (ARS), New York/VG Bild-Kunst, Bonn, Germany. **Plate 18:** *Encyclopedia of the Tanakh* (Bible Encyclopedia). **Plate 19:** ASAP/Kenneth L. Fischer. **Plate 20:** Kelly West.

Glossary of Hebrew and Yiddish Terms

aggadah—nonlegal portion of the Jewish oral law, often in the form of anecdotes, proverbs, homilies, stories, and parables.

aliyah—(1) during prayer services, the calling up of men to read the Torah; and (2) settling in Israel (literally: "going up").

amah—a cubit, from 18 to 23 inches (45–58 cm).

amoraim—sages whose discussions and ruling are recorded in the *Gemara*.

Ashkenazi (pl. **Ashkenazim**)—a Jew whose ancestors came from the Rhine Valley, Germany, or Europe east thereof. Often used as a generic term for Jews of European origin, in contradistinction to those coming from the extended Arab world.

at'halta diGeula—the beginning of the ultimate redemption.

atzilut—a *kabbalistic* concept of the world of God, as He reveals Himself in His abundance.

brokhe—Yiddish for one of the many blessings to be said (from Hebrew, *brakhah*).

dreidel—a spinning top with letters, used during Hanukkah.

eruv (pl. **eruvim**)—the concept of merging of public with private domain or Sabbath with a festival (lit. "merging").

ervah—an erotic stimulus (lit. "pudenda").

etrog—a citron, one of the four species used ceremonially during the festival of Sukkot.

gefilte fish—usually carp, pike, and whitefish, ground up, shaped into balls and cooked. A major contribution of Jewish cooking to world cuisine (lit. "stuffed fish").

Gemara—the record of rabbinical discussions (200–500 C.E.) on the first redaction of Jewish law, the Mishnah.

genizah—place where sacred writing no longer in use is stored before burial.

goy (pl. **goyim**)—used colloquially to refer to a non-Jew. The literal meaning—nation—is neutral, but unfortunately the term as used by Jews at times shades to the pejorative.

haggadah—the text, containing the retelling of the Exodus from Egypt, used to conduct the Passover Seder meal.

halakhah (pl. **halakhot**)—normative Jewish religious law (lit. "the way"). The term is sometimes used as well for any specific law.

halitzah—the ceremony whereby a childless widow is released from the biblical obligation to marry her late husband's brother, entailing removal of the brother's right shoe (lit. "removal of a shoe").

hallah (pl. **hallot**)—braided loaf of bread for Sabbath and holidays (lit. "portion of dough separated").

Hanukkah—holiday commemorating the Maccabee revolt in the second century B.C.E., observed by lighting an additional candle each of eight days. Falls in early winter.

haredi (pl. **haredim**)—a Jew meticulous about observance and religious study, often characterized by reservations vis-à-vis modernity and the state of Israel; "ultraorthodox."

Hashem—a way of referring to the Deity without invoking His name (lit. "the name").

Hasid (pl. **Hasidim**)—a follower of a charismatic Jewish religious movement in Eastern Europe; now often an observant Jew who follows a particular religious leader or "rebbe" (lit. "a pietist").

hillazon—a snail, the source of the biblical blue dye.

hutzpah (Yiddish and Hebrew)—audacity.

Kabbalah—the mystical Jewish tradition.

kashruth—laws governing the permissible and forbidden foods and their preparation.

kebbaneh—a food prepared by Yemenite Jews.

kibbutz (pl. **kibbutzim**)—a collective farm settlement in Israel. A member of a *kibbutz* is colloquially called a *kibbutznik*.

kiddish (or **kiddush**)—the sanctification over the wine, which is recited on Sabbaths and holidays.

kippah—skullcap, sometimes called *yarmulka*.

klaf—parchment from a kosher animal hide.

kosher (Hebrew: **kasher**)—ritually fit; often applied to foods permitted by Jewish law but can also pertain to ritual objects (e.g., a *sukkah*, phylacteries).

lulav—palm branch, one of the four species used during the Sukkot festival.

maneh—a measure of weight equal to 50 *sheqel*, approximately 1 pound.

matzah—unleavened bread, eaten during the Passover festival.

megillah—a scroll; usually refers to the scroll of Esther read on the festival of Purim.

meshuggah (Yiddishism)—insane, madcap.

mezuzah—parchment inscribed with biblical passages in a container affixed to doorpost of a Jewish home or room as a visual reminder of a spiritual obligation (lit. "doorpost").

midrash (pl. **midrashim**)—a statement (story, law, parable, question, etc.) connected to a biblical passage.

mikvah—a specially constructed pool for ritual immersion. Used (a) mostly by women; but (b) also (and separately) by men; and (c) for immersing certain kitchen utensils.

Mishnah—the body of postbiblical Jewish legal material redacted about 200 C.E.

Mishnah Brurah—commentary by Israel Meir HaKohen (d. 1933) on the Shulhan Arukh code of practical, everyday *halakhic* matters.

Mishneh Torah—(a) the fourteen-volume comprehensive code of Maimonides (d. 1204, Egypt); (b) another name for Deuteronomy.

mitzvah (pl. **mitzvot**)—precept, commandment. There are 613 *mitzvot*.

oy vey (Yiddishism)—expression of woe.

pogrom—an organized massacre, especially of Jews.

posek—a rabbi who may render a *halakhic* legal decision (a *pesak*).

Purim—late winter feast day, described in the scroll of Esther.

rabbi, rebbe, rabban, rav, rabbenu (also abbreviated R.)—all these terms are roughly equivalent honorifics, attached to a learned Jew ordained to teach, preach, or decide *halakhic* questions (lit. "master"). Ordination as a rabbi does have to follow certain established procedures, which may vary from one Jewish community to another.

rabossai (Yiddishism)—respected friends, often used jocularly (from the Hebrew: *rabotai)*.

Rashi—preeminent commentator (1040–1105, France); also a common reference to his commentaries, which are often printed alongside the Bible and Talmud.

responsum (pl. **responsa**)—an answer, often by a learned rabbi, to a question about *halakhah*.

rodef—one who pursues to kill or violate another person.

Sanhedrin—supreme religious and judicial body in the land of Israel during the Roman period and until 425 C.E. The smaller Sanhedrin had twenty-three members, the larger seventy-one.

Seder—the Passover service, including the reading of the *haggadah* and a meal.

sefer—a book.

sefer Torah (pl. **sifrei Torah**)—a scroll of Torah (Pentateuch) written on parchment (lit. "a book of Torah").

Sephardi (pl. **Sephardim**)—a Jew whose ancestors came from Spain, the Mediterranean, or Middle Eastern countries (lit. the Hebrew word for Spanish).

shaatnez—forbidden mixture of wool and linen.

Shekhinah—the immanent aspect of the divine (represented as a feminine figure) which went into exile with Israel following the destruction of the Temple.

sheqel—basic biblical weight; also a unit of currency, a coin.

shmatteh shlepper (Yiddish)—rag collector (lit. "rag dragger").

shtetl—small Jewish town of Eastern Europe.

Shulhan Arukh—code par excellence of *halakhah*, written by Joseph Caro (d. 1575 Safed).

sotah—a woman under suspicion of adultery.

streiml—a broad-brimmed, fur-trimmed black hat worn by some *Hasidim*.

sukkah—booth or tabernacle, which Jews are commanded to dwell in during the Sukkot festival.

Sukkot—early autumn festival celebrated by dwelling in tabernacles and making blessings on the four species (palm branch, citron, myrtle, willow).

s'khakh—the thatch or other once-living material covering the *sukkah*.

tahor—ritually pure, not defiled.

tallit—prayer shawl that has a specially knotted fringe in each corner.

Talmud—the major Jewish religious text, other than the Bible (lit. "learning"). Originally transmitted orally, then written down, it consists of the deliberations of generations of rabbis in postbiblical times. It received its final form between 200 and 600 C.E. The Talmud contains the Mishnah and *Gemara,* and in modern printed versions various commentaries are included, such as that of Rashi. There is a Jerusalem Talmud, and a more extensive Babylonian one.

tamei—ritually impure, defiled.

Tanakh—the twenty-four books of Jewish Scripture, the Jewish Bible (Pentateuch, Prophets, Writings).

tannaim—the sages who are responsible for the Mishnah.

tefah—a fist, 3.2–3.8 inches (8–10 cm).

tefillin—phylacteries; small black leather boxes containing biblical passages on parchment, which a male Jew is commanded to put on his forehead and (if he is right-handed) on his left forearm during weekday morning prayers.

tekhelet—the biblical blue dye (and the wool dyed with it) from a secretion of a certain kind of snail (Numbers 15:37).

teva—nature.

tikkun, tikkun olam—the concept of repairing, improving the world around you (lit. "repair").

Torah—the Pentateuch, five books of Moses; in the broader sense, Torah stands for Jewish learning; the way of life as ordained in the Bible and rabbinic sources.

tosafot—twelfth- to fourteenth-century comments on the Talmud by Franco-German scholars beginning with Rashi's descendants.

tush (Yiddish)—a colloquial term for an essential but (to some) unmentionable part of the human anatomy.

tzitzit—the biblically required fringe on each corner of a four-cornered garment (lit. "a fringe").

tzniut—modesty.

yeshivah (pl. **yeshivot**)—A Jewish religious academy, usually for higher Talmudic studies.

yeshivot hesder—*yeshivot* in Israel where young men alternate periods of study with army service.

yekke—endearing, though occasionally derogatory term used for Jews of specifically German origin by other Jews.

Zohar—the primary text of the *Kabbalah*, Jewish mystical writing. Attributed to Moses of Léon, late thirteenth century C.E.

Index

A

Abraham ibn Ezra, 50, 124
Ackerman, Diane, 34, 147–148
Adam, 272; R. Soloveitchik on two Adams, 51–54
aesthetics, 67, 346n8
aggadah, 31, 60, 67; Nahmanides on, 311n9
Ain Hawarrah, 140; map 140
Ain Mussah, map 140; illus. 145
Akiva, Rabbi, 67
Al Andalus, 26, 43; Hebrew poetry in, 25–29
alloys: bronze, 249; of copper, 249, 255; steel, 250–251; strength of, 240, 249
Amichai, Yehuda, 201–203, 208
Amundsen, Roald, 293
Amutat P'til Tekhelet 298, 328n10, Plate 10
Arabs/Arabic 50; coexistence with, 26–29, 208; modesty of, 33
Aristotle, 167
arrow: in the art of Paul Klee, 232–233; as biblical metaphor 218
aspartame, 83
Augustine, St., 134

B

Bader, Alfred, 96–97
Bar Ilan Responsa Project, 68, 112, 312n21
Barrett, Pip, 138
Bartolommeo di Giovanni, 272
Ben-Gurion University, 293, 296, 297
Berger, Michael S., 118–119
Berry, Wendell, 9–10, 42
Bialik, Hayyim Nahman, 286–287
Bible, the, 58–59
Bible passages: Gen. 1:28, 2, 36, 51; 2:4–15, 51; 17:11, 219; 21:30, 219; 31:52, 219; Exod. 15:4, 5; 15:22–25, 123, 153, 156; Lev. 19:2,19, 243; 26:44, 275; Num. 5:17–24, 30, 152; Ch. 15:17, 159–164, 206; Deut. 16:13, 64, 73; 17:8–11, 85; Josh. 10:12, 28; 2 Kings 2:21,152–154; Isa. 5:20, 150; 53:6,13, 304n14; Jer. 6:27–30, 239; Jon. 4:11, 121; Mal. 3:2–3, 246; Ps. 18:31, 245; 23:1–4, 16–18; 24:3–4, 241; 42:2, 16; 115:5–16, 50; Job 38:24–31, 4; illus. 12,15, 83. *See also* Torah.
Bijvoet, Johannes Martin, 96
Biot, Jean-Baptiste, 86, 89–90, 291
bitter waters, 123–157; ordeal of, 31, 151–153, illus. 153; origin of bitterness, 138–145
bitter taste in food, 137
Bleich, J.D., 23–24
blessings: over fringes 160, 194; upon handwashing 206; over miracles, 131–132; mixed blessings, 234; by scribe 283, 347n33; over wine 55
Blidstein, G., 305n20, 308n66
blue pigments, 165–186
blue-dyed Celts, 172
Bonner, William, 107
Book, people of the, 289
books, 263–289
Bracewell, Ronald N., 84
Brann, Ross, 28
Brody, 268
Bruchac, Joseph, 24–25, 54
Buddha, 242
bug-eyed monsters, 107

C

cabbage looper moth, 248, 291
Caillebotte, Gustave, 33, 217
calendrical dispute, 91, 119,
    291
camel caravans, 263, 272
Carmy, Rabbi Shalom, 119,
    205, 303n3, 347n32
Caro, Joseph, 61
cellulose, 21, 281
Chagall, Marc, 104
chaplains/chaplaincy, 266–268
chemistry, definition of, x
Cheyenne legend, 48
chicken soup & *matzah* balls,
    129, 234
chirality/chiral molecules,
    80–84; origins of, 106–109; of
    threads and yarns, 82; illus.
    81–84, 88, 114
chlorofluorocarbons (CFCs),
    44–47
chronology, table of, 69
Chymical Wedding, 253
Coca-Cola, xi, 257–258, 261,
    291
Cohen, Rabbi Shlomo, 48–49
Colette, 240–241
collagen, 273, 276
commentaries on Torah and
    Talmud, 60–61. *See also*
    Abraham ibn Ezra; Hirsch;
    Hizkuni; Nahmanides; Rashi
copper: alloys, 249, 255; vs
    silver, 250
cotton, 11, 19–21, 23, 24, 39,
    281
court case, 213–238
Creation, ix, 2, 4,17, 36, 40, 79,
    109, 129, 135
cupellation, 239–240
cuttlefish, as source of
    tekhelet, 174–185

D

Da Costa, Beatriz, 108
Day, P., 138
Dead Sea Scrolls, 270, 273
Delaney Clause, 257
Dembs, Rabbi Reuben,
    282–283, 289, illus. 289
Devil's Advocate, 133
dietetic sweetener 83
DNA, RNA, 37–39
Domb, Cyril, 119–120, 315n22,
    348n1
Douglas, Mary, 243–244, 254,
    258–261
Dr. Jekyll and Mr. Hyde, 248
dye chemistry, xi, 48–49,
    177–185

E

Eco, Umberto, 221
ecological concerns and
    Judaism, 48–49, 275
Egyptian metallurgy, 239–240,
    Plate 18
Einstein, Albert, 79
Eisner, Thomas, 248
Eleazar of Modiin, Rabbi, 126,
    128
elephants: as kosher *sukkah*
    walls, 66–68, illus. 67; as non-
    kosher food, 66, 327n9
Eliezer ben Hyrkanus, Rabbi,
    110–111
Elisha, 152
Emerson, Ralph Waldo, 17
enantiomers, 81–82; absolute
    configuration of, 96; distin-
    guishing, 93–94
Entebbe, rescue at, 131–132
entropy, 247–248; of Caesar
    salad, 248–249
environmental concerns, 44–50
Ephraim, 98

Epstein, Barak, 40–41, 53–54
ervah, 33–34
Eschenmoser, Albert, 38
ethics, and *halakhah*, 70, 172

F

Fan K'uan, 12–13, Plate 4
Fascism, 260, 261, 262
fathers of superb chemists, 109
Feinstein, Rabbi Moshe,
    118–119; 305n30
Fletcher, S., 138
fraud in science, reasons for
    unimportance, 118
Freehof, Rabbi Solomon B.,
    271

G

Gamaliel, Rabban, 67, 91, 119,
    291
Geffen, Rabbi Tobias, 257–258,
    291
Gemara (Talmud), 60–67, 208;
    on geometry, 70; on headcov-
    ering 31, 33; on incense, 256;
    on professions, 275; semi-
    otics of, 110,113; on torts, x;
    illus. 63, 71
genizah, 269
Gershom, Rabbenu, 92
giants, shoulders of, 115–116
Gittenberger, Edmund,
    120–121
Goldberg, Norman, 139, 145
Goldfinger, Andy, 139
Gómez-Pompa, Arturo, 10
Greenberg, Moshe, 154–155
Guillén, Nicolás, 246–247

H

H., Professor, 115
Haggadah illustration, 12,
    Plate 1

Hair, lyrics from, 34–35

halakhah, xi, 34, 58,60, 111, 119

Halberstam family (Sanz *Hasidim*), 32, 284, illus. 284

Halpern, Moshe Leib, 264

handedness of molecules. *See* chirality

Handel, George Frideric, 13–14, 245

Harvard, 1,17,129

Hasidim: haircovering of, 30–32; in Naples, 174–176; Radzyn, 174–185; Sanz, 30, 32, 282–86

Hayyot, Rabbi Zvi Hirsch, 268–269

Hercules, 168

Herzl, Theodor, 192–203; flag suggestion, 193–201; illus. 193, 196, 200. *See also* Zionist Congress

Herzog, Rabbi Joel, 176–178

Herzog, Rabbi Isaac Halevy, 171–172, 176–178; illus. 171

Hillel, school of, 110

Hirsch, Rabbi Samson Raphael, 235, 313n27

Hizkuni, 129

Hockney, David: *Pearblossom Hwy*, 215–216, 232; *The Splash*, 225

Hoover, Herbert and Lou, 239–240

Huxley, Aldous, 256

I

impurity, and mixtures in religion, 256, 260; power of, 254. *See also* purity

incense, 256

indigo, 166–174; brominated, 169–170; chemical structure of, 169; recipe for making, 173; synthetic, 170; urine in the making of, 173

inks, 280–281

insects, as chemists, 248

Ishmael, Rabbi, 67

Islamic Spain. *See* Al Andalus

Issar, Arie, 145

J

Jacob, 96–100, 299

jeans, 170; Miracle Boost, 151

Jewish Catalogue, 75, 77

John Paul II, Pope, 309n77

Johnston, Harold, 46

Jonson, Ben, 35

Joseph, 97–98, 99, 272

Judah Halevi, 27, 270; hand writing, illus. 270

K

Kabbalah, 41

Kadesh Barnea, 140, 142

Kampen, Martin, 232

Katzenellenbogen, Rabbi Ezekiel, 269

Kaus, Andrea, 10

kebbaneh, 234

Kerr, Clark, 215

Kibbutz Mizra, 204, 332n58

Kiryat Sanz, 30–34, 282–284

Klee, Paul, 232–234; *Conqueror*, Plate 15; *Eros*, Plate 17; *Mask of Fear*, Plate 16

Kolbe, Herrmann, 100–103, 317n35

Kook, Rabbi Zvi Yehuda, 205

Kook, Rabbi Abraham Isaac, 204, 205

Korah, 129, 159–165, 186–192

Korah's wife, 161

Korean War, 266

Kovno, 57–58, 59, 289

Krumbein, Wolfgang E., 143–144

L

laboratory practice, penalty for faulty, 256

Lacaze-Duthiers, Félix Henri de, 168

Langmuir, Irving, 116

Lavoisier, A. L., 292

Le Bel, Joseph Achille, 100

leather, 273–282

"Left Right Out of Your Heart," 106

Leibowitz, Nehama, 136–137, 146, 294, 298

Leibowitz, Elhanan, 294, 298

Leibowitz, Yeshayahu, 294; on miracles, 134

Leiner, Rabbi Gershon Hanokh, 174–185

Lele people, 243

Levi, Primo, 262, 277–279

Lewis, G.N., 116

limonine, 83

Lourdes, miraculous cures at shrine of, 133

Luke, Gospel of, 284–285

Luria, Rabbi Yitzhak, 105

M

Maga, J. A., 137

magnesium salts, 139–140

Maimonides 22, 61, 68, 103, 205, 257; on Purities 339n24; on qualities of judges, 222; on self-control, 257; on *sukkah*, 68; on symbols, 236

Manasseh (Manasheh), 98

Marah, 123–157, 186; location of, 140–145

Marrakesh, 271–272

Masada, 31

matzah balls, 234
Meir, Rabbi, 172
Menger, Fred, 141–142
Messas, Rabbi Joseph, 271
midrash, 60; on aesthetics; on bitter waters, 126–127, 143; on Creation, ix; on Korah, 160–65, 187–188, 327n4; on purity, 245; on *sukkah*, 67; on truth, ix
mikvah, 224, 283; illus. 224
miracles, 123–157; in Catholicism, 132–133
mirror images, 80–83
Mishnah, 62–65; on environment, 48, 275; on headcovering, 30; on *sukkah* 64, 73; on miracles 129
mitzvah, mitzvot, 257, 288, 310n4; importance of, 189–192; relating to nature, 11; of procreation, 2, 51, 53; settling Land of Israel, 205; *sukkah*, 56–58; *tzitzit*, 22–23, 189. *See also* aesthetics; blessings
mixtures, 239–262; animal breeding, 244; geographic, 244 ; marriage, 244; meat/milk, 242, 258; metaphysics, 244–245; in religion, 256–260. *See also shaatnez*
molecules and chemical structures, x, 20–22; CFCs, 45; creation of, 37–40; dibromoindigo, 169; icosahedron, 36; Prussian blue, 183; sanctity of, 282; soccer-ball-shaped, 36; superconductors, 254; water, 227–229; ziziphin, 149; illus. 81, 82, 84, 101, Plate 7. *See also* chirality; natural/unnatural
Molina, Mario, 47
Morell, Sima, illus. 65
Moringa tree, 144

morphine, 83
Morton, Thomas, 255
Moses ibn Ezra, 28–29
Moses, 15, 123–157, 159–165
murex snails, 167–169, 298 Plate 9

N
Nahmanides, 61, 90–91, 103, 112, 174, 205, 218–238, 298; and disputation at Barcelona, 174, 311n9; on interpretation of signs, 218–220; on miracles, 135–136; on rainbows, 218–219, 298
Natan (Nathan), Rabbi, 111, 127, 129
natural/synthetic: polymers, 19–23; nucleic acids, 37–40
natural/unnatural, 76–77, 293, 297
nature and natural: in advertisements, 5–6; in art, 11–15, 33; definitions, 3, 8; domination by man of, 2, 43–44, 50–53; Greek and Latin etymology, 8; Hebrew word (*teva*) for, 5, 7; illus. 6, Plates 1–7
Navajo, 1, 2, 5,18, 49, 286, Plate 20
Nazism, 57–58, 263–266, 289
Netanya, 282–284; 310n7
Neusner, Jacob, 287
nucleic acids, 37–40
nylon, 11, 19–21, 39

O
Ockham's Razor, 117
Offenbach (city), 265–266
old wine, new flasks, 162, 185, 284
old wine, new wineskins, 285
Old Blood and Guts. *See* Patton
oleander, 127, 322n10
olestra, 18, 37

olives, 127
Onn, 162–163
Oppenheim, Moritz, 55–56, 72
optically active molecules. *See* enantiomers
oral law, 59
ordeal of bitter waters, 31, 151–153
Oshry, Rabbi Ephraim, 57–58, 62, 68, 288–290
ozone depletion, 45–47

P
paper, 281–282
parchment, 273–284
parking signs, 213–238; color of, 221
Pasteur, Louis, 86–90, 106, 291
pattern recognition, 237–238
Patton, General George, 266
Peirce, Charles S., 231–232
Pentagon, the, 266–268, 269, 271
pentoses, hexoses, 38
pheromones, 248
photosynthesis, 247
phylacteries. *See tefillin*
pi (π), talmudic estimate of, 72
Pliny, 172
poetry, 24–25, 27–29, 34–35, 83–84, 226–27, 246–47, 281, 286–87; Hebrew in Iberia, 26. *See also* biblical passages (Psalms)
polarized light, 86; rotation of plane of, 86–88
pollution, Jewish attitudes toward, 48–49
polygyny, 92
polymers, 19–22
Potter, Paulus, 12, Plate 3
Prussian blue, 99–100, 177–185, 299
Psalms, interpretations of, 16–17, 44, 50

publication in science, 116–118

purity in science, 255–256

purity/impurity, 239–262; intentionality in, 257–258; Jewish concepts of, 74, 244–245

R

Rabin, Yitzhak, 210–211

racemic acid. *See* tartaric acid

Radzyn *Hasidim*, 174–185

rags, 281

Rambam. *See* Maimonides

Ramban. *See* Nahmanides

Rashi, 61, 68, 78, 85–86, 90, 96, 103, 172

Rav, the. *See* Soloveitchik, Rabbi Joseph B.

Rava, 113

rayon, 19, 21, 39

reading upside-down, 237

representations, of water, 223–229. *See also* symbols

responsa, 48–49, 57–58, 61–62; on dyeing, 61–62; on *sukkah*, 68, 105–106, 118; on *tefillin*, 105; on Torah scrolls, 267–73. *See also* Oshry, Rabbi E.

rhetoric, 261

right- and left-handed molecules: distinguishing, 93–94; origins of, 106–109. *See also* enantiomers

Rishon Le-Zion, 197–198, 207, Plate 11

rodef, 57, 207–211

Roelofs, Wendell L., 248, 256

rose oil, 247

Rosenberg, Alfred, 265–266

Rosh, the, 112

Rouseff, Russell L., 137

Rowland, F. Sherwood, 47

royal purple. *See* Tyrian purple

S

sabkha, 143–144

Sabkhet el Bardawil, 140, map 140

saffron, 279

Safran, Hilel, 263

saints, 132–133

Sanhedrin, The 90–91

Santa Sabina, Basilica, 15

saponins, 149–150

Saussure, Ferdinand de, 231–232

Scheindlin, Raymond, 28–29

scholars: talmudic penalties for, 209; respect for, 112, 215; disagreement among, 113

Schurtz, C. N. illus., 153

scribe, work of, 282–283

semiotics, 213–238; of Talmud page 110, 113

settlers in Judaea and Samaria, 203–211

sex: and halakhah 288; of tekhelet snails, 170. *See also ervah*; pheromones

shaatnez, 243, 257, Plate 19

Shamai, school of, 110

Shaw, George Bernard, 130

Shenkar College, 179

sheqel, 196–197

Shimon ben Gamaliel, Rabbi, 126

Ship Rock, 2, 286, Plate 20

Shulhan Arukh, 238; on miracles, 131

signs, 9; iconic, 231–232; semiotics of, 219–220; symbolic, 231–232; traffic, 230, 213–238

Silberstein, Rabbi Eli, 299, 318n37

silver vs copper, 250

snails, right- and left-handed, 120

Solomon ibn Gabirol, 27–29

Soloveitchik, Rabbi Joseph B.,

118–119, 189–192; story of two Adams, 51–54

sotah, 151–153. *See also* bitter waters

Spanier, Ehud, 170, 297

Steinsaltz, Rabbi Adin, 74

Stevenson, Robert Louis, 248

Stoltzenberg, Daniel Stoltzius von, 253

Strauss, Richard, 14, 22, 43

sublime, 16–18

sukkah/sukkot, 1, 55–78, 293–4, 297; codes on, 68; collapsing, 78; commentaries on, 68; dimensions of, 64; elephants and, 66–67; *Gemara* on, 64–66; materials science of, 62–69; mathematics and, 70–71; *midrash* on, 67; Mishnah on, 63–64; and the natural, 72–78; responsa on, 57–58; Talmud on, 62–67; walls of, 64–66, illus. 56, 65, 67; Plate 8

Sukkot, Festival of, 55–78, 104

Sumerian disputation, 250

superconductors, 254

symbolism, 298. *See also* colors; Hirsch, S.R.; signs; *tekhelet*

synthesis, chemical, 36–40

synthetic fibers, 19–22. *See also* natural/synthetic

s'khakh (roofing for *sukkah*), 72; requirements for, 73–75

T

Tabernacles, Festival of. *See* Sukkot

Tacitus, 260

Tales of Hoffmann, 293, 297

tallit, 194–195, illus. 59, 104, 160, Plate 11. *See also* tzitzit

Talmud, 60, 62–67; debates in, 110; illustration of page of, 63; Jerusalem, 90. *See also* Gemara

Tamir, Abraham, 293, 297

tanning, 278–279. *See also*
leather

tannins, 274

tartaric acid, salts of, 87–89

taste: biology of, 147–149;
modifiers, 147–149

tefillin, 67, 104–106; illus. 104,
160

tekhelet, 159–212, 297, 298,
Plates 9–12; color of, 172;
forgeries of, 172; symbolic
meaning of, 189–192. *See also*
Amutat P'til Tekhelet; Leiner;
Rabbi H.; and Ziderman, I. I.

teva (nature), origin of word,
5, 7

tikkun olam, concept of, 42,
49, 308n66

time-line of people and events,
69

Torah, 58–59, 267–268, 310n7;
dual, 58–62; installation of,
283–284; scrolls, 282–283;
illus. 59, 267, 284

Tournier, Michel, 261

Tu'Bshvat, 11, 124

Tyre, 167–169

Tyrian purple, 166–174

tzitzit, 159–165, 194; materials
for, 22–23; 305n28; illus. 160.
See also tallit

tzniut, concept of, 31

U
U.S. Armed Forces, 266–268

V
van der Werff, Adriaen,
96–100, 299

van't Hoff, Jacobus Hendricus,
100–103

vellum, 273–282

Vermeulen, Jaap J., 120

W
water: as source of letter M,
227; chemical structure of,
227–229; composition of, 247;
representations of, 223–229

Weininger, Stephen J., 96

Weiss, Rabbi Yitzchak, 105

White, Lynn, Jr., 46

wigs, 30–34, 297, 308n59–60,
Plates 5, 6

willow, 126, 141

Wilson, E.O., 24

Wilson, Ken, 117

Wislicenus, Johannes, 102–103

woad, 172–173

Wölfflin, Heinrich, 98–99

Wolffsohn, David, 192–201

Woodward, R.B., 117, 281

Woodward-Hoffmann rules, 15

Y
Yehiel Mikhal of Zloczow, 264,
291, 292

Yehoshua ben Hananiah,
Rabbi, 91, 111, 119, 126, 128,
291, 319n52, 324n33

Yehoshua ben Korha, Rabbi,
127, 128

Yemenite Jewry, 234–238; illus.
237

yeshivah, 1, 70; and army joint
program (*hesder*), 113, 206,
320n59, 332n1, 332n57

Z
Ziderman, Irving Israel,
16, 30, 34, 328n14,

Zionist Congress, First,
192–201, 210

ziziphin, 149–150

Zloczow, 263–265, 268, 292

Zohar, 152, 275